CURVE STITCHING

The art of sewing beautiful mathematical patterns

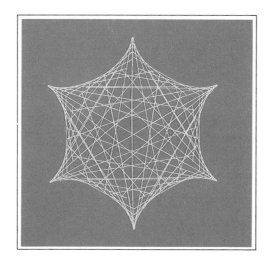

Jon Millington

Tarquin Publications

"In my young days cards of different shapes were sold in pairs, in fancy shops, for making needle-books and pin-cushions. The cards were intended to be painted on; and there was a row of holes round the edge by which twin cards were to be sewn together. As I could not paint, it got itself somehow suggested to me that I might decorate the cards by lacing silk threads across the blank spaces by means of the holes. When I was tired of so lacing that the threads crossed in the centre and covered the whole card, it occurred to me to vary the amusement by passing the thread from each hole to one not exactly opposite to it, thus leaving a space in the middle. I can feel now the delight with which I discovered that the little blank space so left in the middle of the card was bounded by a symmetrical curve made up of a tiny bit of each of my straight silk lines."

So wrote Mary Boole in 1904, some sixty years after she had invented curve stitching.

Acknowledgements

I should like to express my gratitude to my wife for her invaluable advice and encouragement. The original idea for this book was hers and, together with Jane, Libby and Susan, she sewed many of the patterns.

My thanks go also to the pupils of Clifton College Preparatory School for sharing my interest in curve stitching, and to John Barrett for his help with the computer program printouts.

J. M. M.

The works by Naum Gabo, illustrated on pages 86 and 87 have been reproduced by kind permission of Nina and Graham Williams. The work by Antoine Pevsner illustrated on page 87 has been reproduced by kind permission of ADAGP Paris and DACS London 1989. The works illustrated on page 88 have been reproduced by kind permission of The Henry Moore Foundation. The works by Barbara Hepworth, illustrated on page 89 are reproduced by kind permission of Sir Alan Bowness.

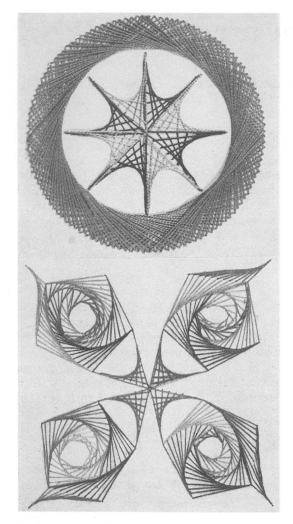

From "A Rhythmic Approach to Mathematics" by Edith L. Somervell, first published in 1906.

© 1989   : Jon Millington
I.S.B.N   : 0 906212 65 0
Editor   : Magdalen Bear
Design   : Philip Streeting
Printing   : Loxley Brothers Ltd. Sheffield

Tarquin Publications
Stradbroke
Diss
Norfolk IP21 5JP
England

The art of sewing beautiful mathematical patterns

Curve stitching is a topic which can be approached on many different levels. First and foremost it is a handicraft which is accessible to quite young children and which can be enjoyed without any need to understand the mathematical fundamentals which lie beneath it. At this level it is best to avoid the need for too much accurate measuring. Lovely patterns can be made simply by marking equal lengths along straight lines using a ruler. For circle designs, a protractor or a simple template can be used to mark equally spaced points around a circumference. From such simple beginnings many of the designs in this book are generated. As skill and enthusiasm develops it is natural to progress to polygons and more complex patterns.

The computer offers another line of approach. It is an excellent tool for investigating the possibilities of curve stitching. In this book there are 35 programs written in both BBC and Spectrum Basic. None of the programs is very long and all may be modified at will. With some knowledge of programming they can be adapted for other computers. The illustrations in the computer section also form a useful index of possible designs for curve stitching, whether or not a computer is available.

At a deeper level, we may think of curve stitching as an essentially mathematical activity. Why for instance does a parabola arise when sewing between equally spaced points on a pair of straight lines? Can we generate other curves so simply? What mathematical properties does a chosen curve have? Would one of those properties allow us to generate the chosen curve by a simple sewing rule? It is very satisfying to be able to sew a cardioid from equally spaced points around a circle by using such a rule. But since a cardioid is an epicycloid, can other epicycloids and hypocycloids be similarly generated? Such investigations are the very stuff of mathematics and it is hoped that this book will stimulate such endeavours.

However, let us never lose sight of the fact that curve stitching is a practical activity and that the test of every method is that it really works and is pleasing and satisfying to sew.

Proportional scale for equal divisions along a straight line

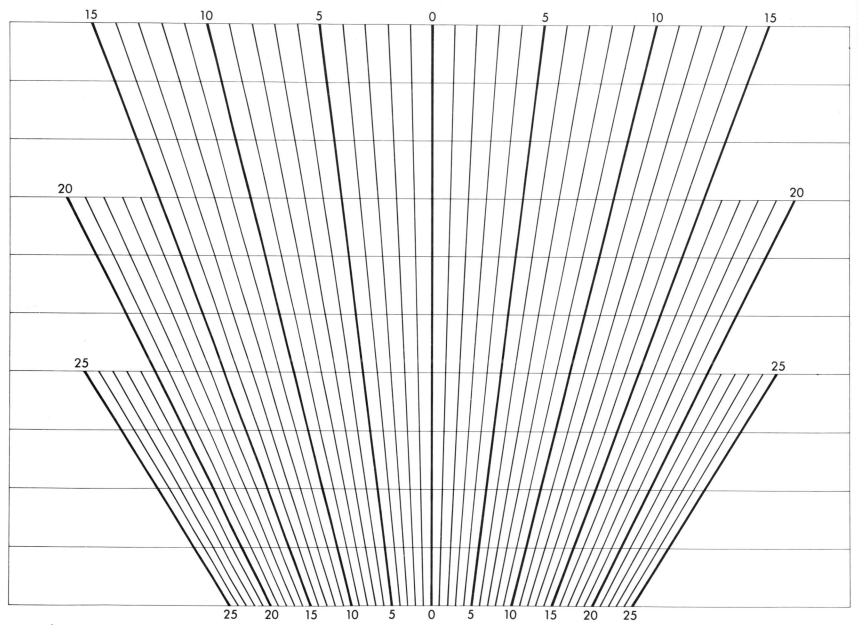

1.
Getting started

Getting started

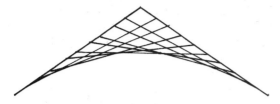

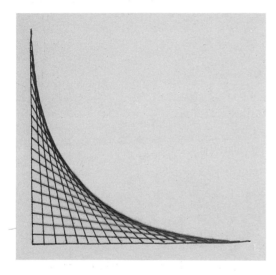

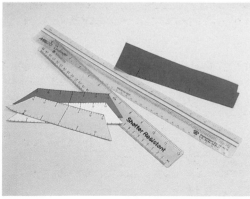

It is remarkable that a curve can be suggested by comparatively few lines, each contributing a very small part to what is known as the envelope of that curve.

Often it is a good idea to draw any intended pattern before starting to sew. In this way you can decide the best overall size and the spacing to adopt between points in the sewn version. Moreover, a drawing will possess its own attraction whether done in pencil, ink or crayon.

The best examples of curve stitching occur when the number and spacing of the threads seems "about right". Exactly what is meant by this expression is an artistic rather than a mathematical judgement, but one that most people seem to understand after only a little practice.

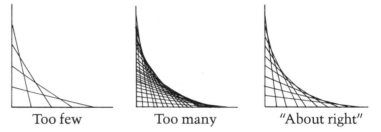

| Too few | Too many | "About right" |

Although a trial drawing may be done to get an impression of what the front of the finished design will look like, all the marking out for the actual sewing is done on the back of the card. Only the points where the holes are needed are pricked through and they should be as small as possible. Use the smallest needle which can be threaded with the chosen thread. Holes which are too large are conspicuous and detract from the elegance of the finished design. Likewise every effort should be made to make sure that no pencil marks or any writing or dirty marks appear on the front of the card.

Marking out straight lines

The most obvious way of marking equal divisions along straight lines ready for pricking through is to use a pencil and ruler. If the ruler has both metric and Imperial markings then it will be easier to find a suitable spacing for the design you have in mind. Alternatively, you can use the templates on page 93 or cut out strips of card and make your own templates using the proportional scale on page 4. As long as the points on a line are evenly spaced the exact spacing is not critical and you will have no difficulty in arriving at something suitable.

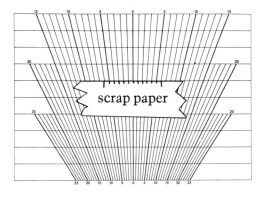

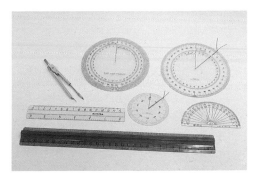

Using a proportional scale

Quite often, however, you will not be starting from scratch, but will need to fit a certain number of divisions into a fixed length. Perhaps you want to divide the side of a polygon into 12 equal lengths. Fortunately, if none of the fixed scales would work out exactly, there is a very simple solution to the problem. With the proportional scale on page 4 you can divide any length into any number of equal divisions. All you need is a straight-edged piece of scrap paper.

On the straight edge of the piece of scrap paper, mark off the distance you want to divide into 12. Place it on the proportional scale and slide it up and down (keeping it parallel to the horizontal lines) until there are 12 equal spaces between your marks. Then copy these divisions on to the edge of the scrap paper and use it as a template to mark off the line on your design.

Marking out circles

Circles drawn with a pair of compasses can be of any convenient size. Using a protractor to measure angles, you can divide the circumference into the required number of equal lengths. It is always easier to work with whole numbers of degrees which are factors of 360° and so certain numbers of points are much easier to mark than others, as shown in the table.

No. of points	6	9	10	12	15	18	20	24	30	36	40	60	72
Angle	60°	40°	36°	30°	24°	20°	18°	15°	12°	10°	9°	6°	5°

A circle of radius 52mm is a suitable size because many protractors are just over 100mm in diameter and so will just fit inside the circle, enabling its circumference to be marked out easily. The extra 2mm allows for the thickness of the pencil line. This is a convenient size for a great many designs.

An alternative method to using a protractor, and one which many people find easier, is to use one of the circular templates from the back of the book. The 36 point template could be used to mark out 72, 36, 24, 18, 12, 9, 6, 4 or 3 point circles and the 40 point template to mark out circles with 80, 40, 20, 16, 10, 8, 5 or 4 points.

The radius of each of the templates is 65mm with a half-size circle inside. This is a convenient size and it was used for many of the examples of curve stitching illustrated in this book. If you want to make templates of other sizes, then use the proportional scales on pages 91 and 92 to construct them.

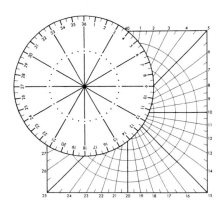

Using mechanical assistance

Another way of avoiding drawing and measuring is to use a gear-wheel from a construction set or a redundant piece of machinery. You can then mark out one point for each tooth. Meccano part number 95 is a sprocket wheel with 36 teeth and there is a Spirograph wheel with 72 teeth. By marking alternate teeth, or indeed one tooth in 3 or 4, you can divide circles into fewer equal divisions. Of course, multiples of 36 are not essential. It depends on what size gear-wheels you are able to find.

Marking out regular polygons

Many patterns make use of regular polygons such as squares, hexagons, octagons etc. Each of them can be constructed within a circle. Use a protractor and the table below to mark the points around the circumference. Don't forget that a hexagon is best constructed with compasses alone.

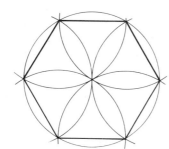

Hexagon

POLYGON	TRIANGLE	SQUARE	PENTAGON	HEXAGON	OCTAGON	NONAGON	DECAGON	DODECAGON
SIDES	3	4	5	6	8	9	10	12
ANGLE AT CENTRE	120°	90°	72°	60°	45°	40°	36°	30°

The templates on pages 93 and 95 also offer a very easy way to construct regular polygons. An equilateral triangle can be drawn with the 36 point template by marking points 12, 24 and 36. To draw a square, mark points 9, 18, 27, 36 on the 36 point template or the points 10, 20, 30, 40 on the 40 point template. To draw a regular pentagon you need to use the 40 point template, a regular hexagon the 36 point template, and so on. With only a little practice you will find this a very quick and easy method.

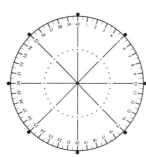

Square

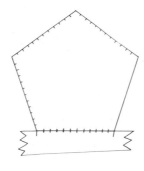

Corners of an Octagon

In order to divide the sides of a polygon into equal divisions the proportional scale is particularly useful.

Materials

Success with curve stitching depends on the balance of the two materials, card and thread. It is the tension of the thread as it crosses the front of the card which gives the straight lines. Unlike embroidery, which is usually done with short stitches on a pliable backing material, mathematical curve stitching is done with long stitches on a firm backing. The card must be stiff enough not to warp under the tension of the thread, nor to tear at the holes, but not be so thick that it is difficult to pierce with the needle. The recommended material is 0.5–0.75mm thick (500–750 micron).

For sewing, ordinary cotton thread works well and is available in an enormous number of colours. It is not liable to build up into a knot as you pull it through the card as is sometimes the case with polyester or artificial silk. On larger designs the thicker polyester button thread, although expensive, is excellent and comes in a wide range of colours.

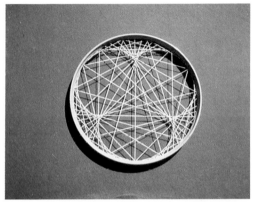

For larger patterns, what about using an old bicycle wheel with the spokes removed? If there are not enough holes you can drill more, perhaps using a metal template to ensure that the extra holes are equally spaced. Make the design itself with coloured string which will look most effective when the completed rim is placed against a contrasting background.

Sewing techniques

Begin by sewing from the back of the card holding the starting end of the thread firmly in position with a paper clip until the design is complete. Then it is often possible to tie the starting and finishing ends to each other to secure both, but if not finish off both ends by sewing them each into their adjacent stitches.

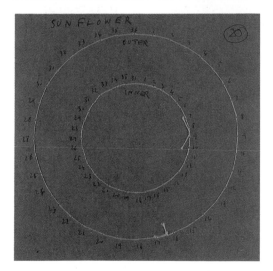

Part of the satisfaction of curve stitching is to waste as little thread as possible and to make the back of the card look very neat. A golden rule is to try to sew across the back only as far as an adjacent hole. Sometimes the choice of an odd or even number of points makes a lot of difference to the neatness of the sewing.

Since it is seldom possible to judge exactly how long a piece of thread will be needed, it is best to master the art of tying a reef knot.

Of course, the knots must always be on the back!

Practical Suggestions

When you have mastered the art of sewing beautiful curve stitching patterns you can frame them as pictures for your walls or desk. Or they could be sewn to make bookmarks or birthday and Christmas cards. Some designs would make a fine decoration on an address or telephone book or could be sealed in plastic to make a set of table mats.

2.
Designs made from parabolas

How to sew a parabola

Since the parabola is the simplest curve to sew, it is a good idea to start with it. It is also highly satisfying.

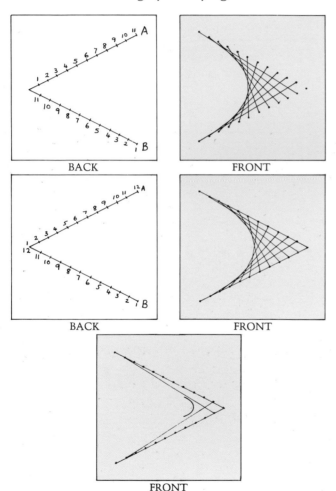

BACK   FRONT

BACK   FRONT

FRONT

Remember that all marking out is done on the back of the card. Start by drawing a pair of lines at any angle on the back. It is better to let them intersect on the card although this is not essential.

Having marked the lines A and B, divide each line into 11 equal divisions numbering them as shown. The intersection is not numbered as it is not going to be used on this occasion. Prick through the points to the front of the card.

Start by holding your thread to the back of the card with a paperclip. Then sew across the front from A1 to B1, B2 to A2, A3 to B3, B4 to A4 and so on, finishing with A11 to B11. Then tie the pair of loose ends together.

If you wish to sew right into the intersection renumber your original diagram as shown. The point of intersection then has the numbers A1 and B12. To avoid covering the holes you have pricked through, start by sewing A2–B2 and continue up to A12–B12, then sew beneath the card to B1 and over the front to A1. Now tie the pair of loose ends together.

It is much simpler to sew a parabola than to explain in words what to do. After you have done a few, you will find that there is no need to number the points. Then you will begin to feel confident about sewing parabolas between different pairs of lines in more complicated designs. In those which follow, this symbol is used to indicate the position of a standard parabola between two lines.

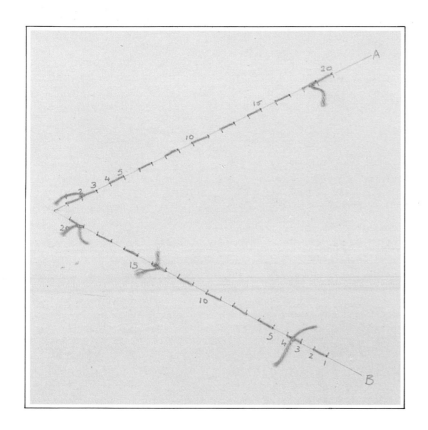

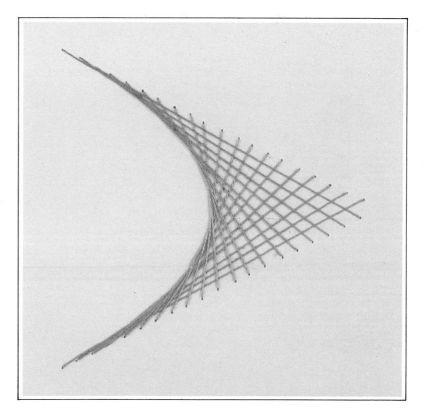

On the back of this parabola, the only stitches showing lie alternately between pairs of holes. As you will never be able to judge exactly how much thread is needed, there may well be several knots. This does not matter as long as they are neatly tied and kept to the back of the card.

You may, as in this case, be able to tie the two ends of the thread together when the sewing is finished. If that is possible, then it gives a neater appearance. If not, scw each end separately into the next stitch or join the ends with a long knot.

This design did not use the point of intersection and so the lines on the back are suggested only by the ends of the stitches. If you then decide that it would look better with the lines clearly defined, it is a simple matter to sew the additional two stitches.

13

1. A pair of opposite parabolas

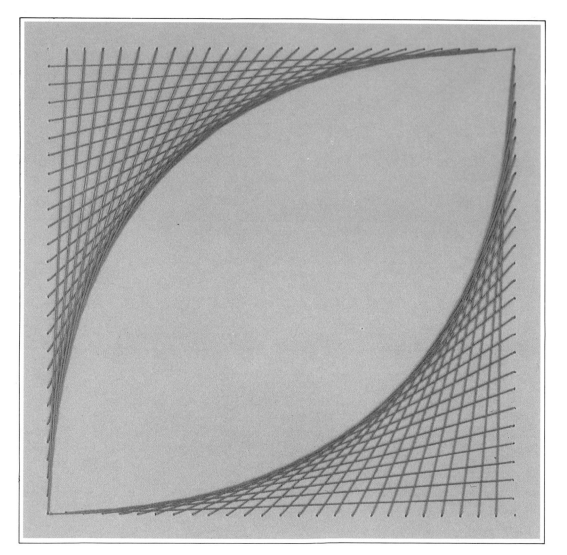

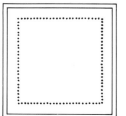

Draw a square and divide each side into 26 equal divisions.

Construct a parabola in the standard way on each of two pairs of adjacent sides.

Notice that here the outline of the square is not sewn in, but is suggested by the ends of the stitches. The holes which mark two of the corners have not been used.

Of course the design could be modified to include the outline and to use these two corners.

2. Zig-zag parabolas

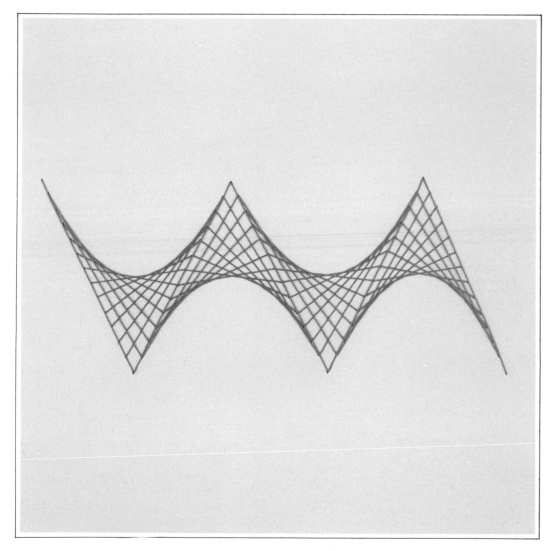

Draw a zig-zag with 5 equal arms and divide each into 13 equal divisions.

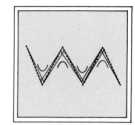

Construct a parabola in the standard way on each of the four pairs of adjacent lines.

This simple method of linking parabolas lends itself to any number of variations. The lines can be at any angle and need not be all the same length.

Do remember that if you start with, say, 12 equal divisions, then each of the other lines, whatever their lengths must also each be divided into 12 equal divisions.

15

3. Three parabolas in a triangle

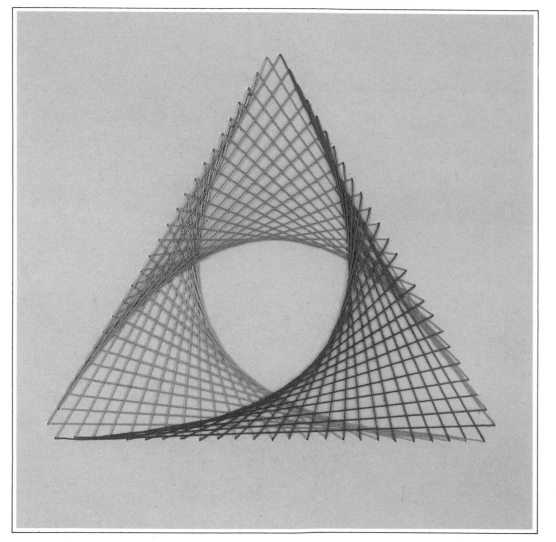

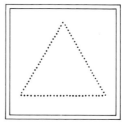

Draw an equilateral triangle and divide each side into 26 equal divisions.

On each pair of lines sew a parabola in the standard way.

Notice how the three parabolas form an attractive shape in the middle of the equilateral triangle.

When choosing the colours, do consider which parabola passes under the other two and which stays on top.

In this design the holes at the three corners are not used; if a sharp outline is preferred they could easily be sewn.

4. Four parabolas in a square

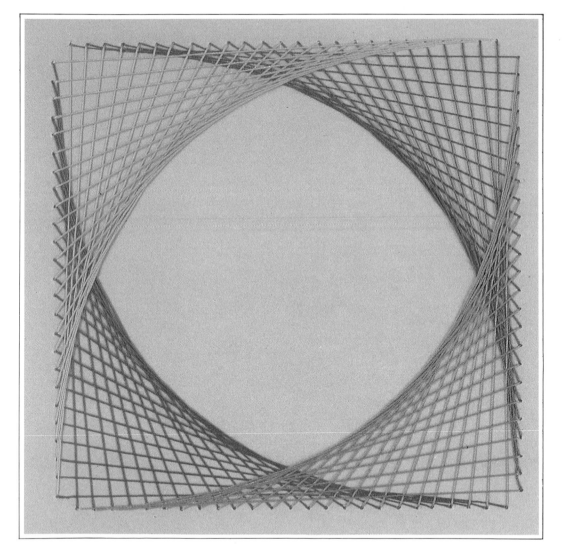

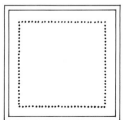

Draw a square and divide each side into 26 equal divisions.

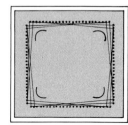

On each pair of lines sew a parabola in the standard way.

This pattern uses four different colours for the four parabolas, but alternatively it might seem natural to make it using only two colours, one for the pair of parabolas which pass under and one for the pair which pass over. Think about which colours lie on top of your finished design as the top colours inevitably partially mask the colours underneath. Again notice that although the actual corners were not used they could have been.

5. Six parabolas in a hexagon

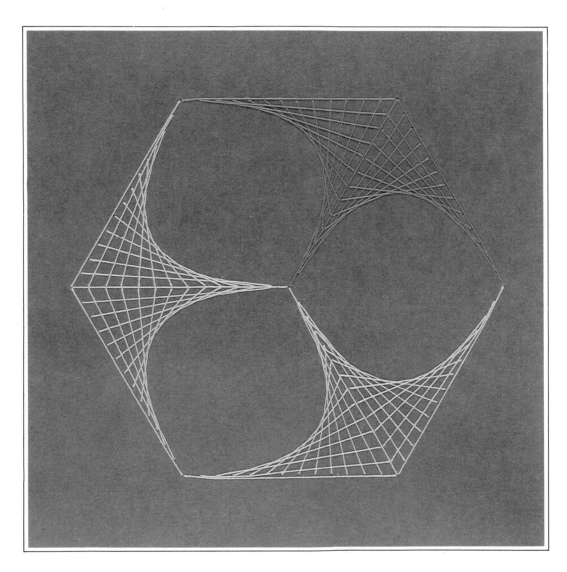

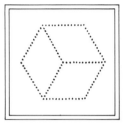

Draw a regular hexagon with 3 radii, making 3 linked rhombuses. Divide each of the 9 lines into 13 equal divisions.

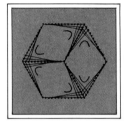

In each rhombus sew two opposite parabolas in the standard way, each time joining a side to a radius.

This pattern gains some of its harmony by making use of the fact that the sides of a hexagon are equal to its radius. A related, but very different design can be made by sewing the parabolas between adjacent sides and adjacent radii, rather than between a side and a radius.

6. Eight parabolas in an octagon

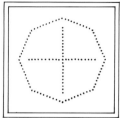

Draw a regular octagon and 4 radii to form 4 equal linked quadrilaterals. Divide each side of the octagon into 10 equal divisions, and then using the same number and length of division mark the radii from the centre.

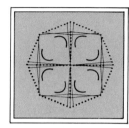

In each quadrilateral sew two opposite parabolas in the standard way, each time joining a side to a radius.

Unlike the hexagon opposite, the radii of an octagon are not the same length as its sides. The radii are longer and so there is a gap on each radius. A simple variation would be to mark out the radii from the corners rather than from the centre, leaving all the gaps at the centre.

More than one colour could also be used but think carefully about effects created when using two, three or four colours.

7. Four pairs of parabolas

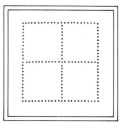

Draw a square and join the centre of each side to form 4 linked squares. Divide each of the 12 lines into 13 equal divisions.

Sew four parabolas in the standard way around the centre of the pattern, and then four more, one at each corner.

This pattern is made from four pairs of opposite parabolas. Ignoring these colours it has mirror symmetry about four axes. Arranging the colours in a different way can create a very different impression and it is a pattern which is well worth drawing in coloured pencils or crayons before starting to sew.

8. Eight pointed star

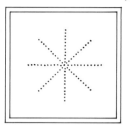

Draw 8 equally spaced radii and divide each radius into 11 equal divisions.

Between each pair of radii construct a parabola in the standard way.

It is probably better to complete the four parabolas in the first colour, before starting on the second.

Notice how the centre point is not used here and how this gives rise to an attractive central star shape. The radii themselves have not been stitched but if they had been they would have intersected at the centre.

The four central parabolas in the illustration opposite are a variation on the same design. It is also possible to make a similar attractive star with 3, 5, 6, 7, 9 or more points.

9. Rotational symmetry from four parabolas

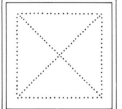

Draw a square and its diagonals. Divide each side into 20 divisions and each diagonal into 36.

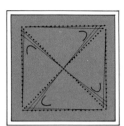

Sew four parabolas in the standard way, each time joining a side and an adjacent half diagonal.

Notice that the diagonals are indicated only by the ends of the stitches, but that the sides of the square are sewn, except for one division.

The windmill design in this illustration looks as if it is being blown clockwise. If each half diagonal had been joined to the other adjacent side the design would look as if it was being blown anti-clockwise.
Superimposing the two versions is yet another idea.

3.
Designs made from circles

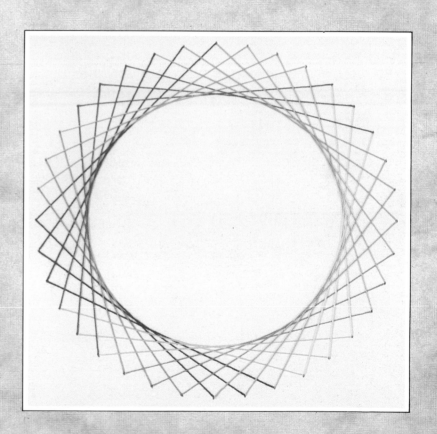

How to sew circle designs

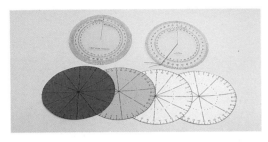

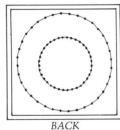

FRONT *BACK*

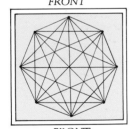
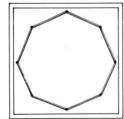

FRONT *BACK*

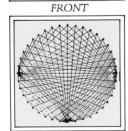
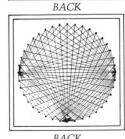

FRONT *BACK*

Many circle designs are even easier to sew than those based on the parabola, once you have divided the circumference into the required number of divisions. The most obvious method to use to mark out the circle is to draw it with compasses and then to measure out the angles with a protractor. However, since certain numbers of points are more useful than others and are frequently needed, it may well be more convenient to make a collection of circular templates. At the back of the book there are two templates which will meet most needs. They can by cut out and stored in a glued pocket inside the back cover. If you prefer to work with different sizes or numbers of points or simply prefer not to cut the book, then there are some proportional scales on pages 91 and 92 which will make it easy to construct any template you are ever likely to need. More detail about marking out is given on pages 7 and 8.

A surprising thing about circle designs is how little thread is wasted at the back. Most only require a stitch to an adjacent hole in order to start the next thread across the front. It is part of the satisfaction of good curve stitching that the front can be so complicated and yet the back so simple.

In this section there are four different mystic rose designs, two complete and two incomplete. For all four the front is both beautiful and elaborate. For the two designs which are complete, meaning that the mathematical concept is fully worked out, the backs are very simple. Stitches only go to an adjacent hole. For the two incomplete designs, however, the backs are as complicated as the fronts.

1. Equal chords

Mark out a 36 point circle.

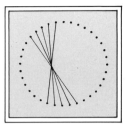

Sew from each point to a point 15 steps further round the circle.

Sewing equal chords within one circle generates another circle which is smaller and concentric. The longer the chords of the larger circle, the smaller the circle at the centre. Indeed, if you sew diameters, then the inner circle disappears altogether. In this illustration, obtained by counting round 15 points, the chords generate a circle with about one quarter of the radius of the large circle.

Patterns of this type are also interesting to draw, as well as sew. When drawing them, other mathematical investigations arise. For example, whether a particular design is unicursal or not. This is investigated on page 76.

2. Two sets of equal chords

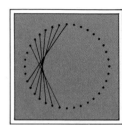

Mark out a 36 point circle.

Using the first colour, sew from each point to a point 11 steps further round the circle.

Then change colour and continue by sewing from each point to a point 15 steps further round the circle.

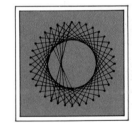

Each hole is used twice for each set of chords and all stitches on the back only reach as far as the adjacent hole. The threads which form the smaller circle lie on top, but this could be reversed to introduce a small change to the design. To introduce a larger change you could sew a third set of chords on top of both.

3. Nine point mystic rose

Mark out a 9 point circle.

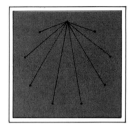

Join every point to each of the other points.

It is probably best to sew all the longest threads first, then the next longest etc, etc. On the back, all threads cross only as far as an adjacent hole.

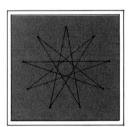

This pattern has a regular nonagon at its centre, a repeat of the shape from which it started. Check that each hole has been used exactly 8 times as it is only too easy to miss some of the diagonals.

It is also interesting to investigate the number patterns arising in mystic roses. Consider the number of diagonals, intersections and regions produced when starting with different numbers of points on the circle. Things are not always as simple as they appear to be initially.

4. Twelve point mystic rose

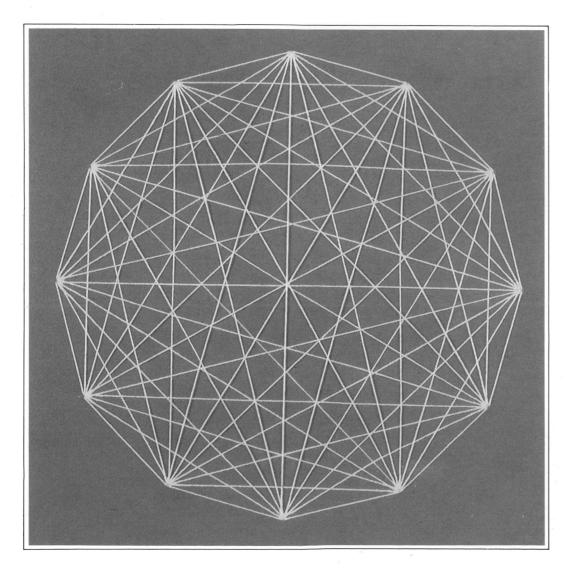

Mark out a 12 point circle.

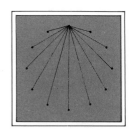

Join every point to each of the other points.

It is probably best to sew all the longest threads first. No threads cross the back but only go to the adjacent hole.

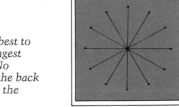

Compare this pattern with the nine point mystic rose. With an even number of points diameters cross at the centre, while an odd number of points leaves a regular polygon at the centre.

Check carefully to see that the design is complete. Each hole is used 11 times, which is one less than the total number of points on the circle. There are diagonals of six different lengths at each hole.

5. An incomplete mystic rose

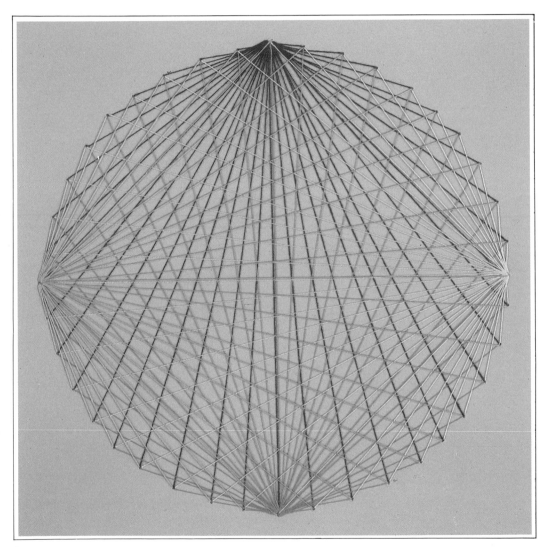

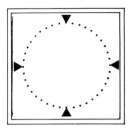

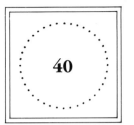

Mark out a 40 point circle.

Starting at the top, mark 4 equally spaced points.

Sew from each marked point to every other point. The back of this design looks very much like the front.

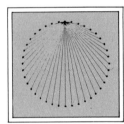

A complete mystic rose which includes all the diagonals from 40 points would be so overcrowded as to lose its attractiveness. However, by choosing only four points and sewing all the diagonals from each in a different colour produces a lovely result.

The order in which the colours are sewn makes quite a difference to the look of the final pattern. Here the blue and red which are the strongest colours were sewn before the orange and yellow.

6. Sunrise

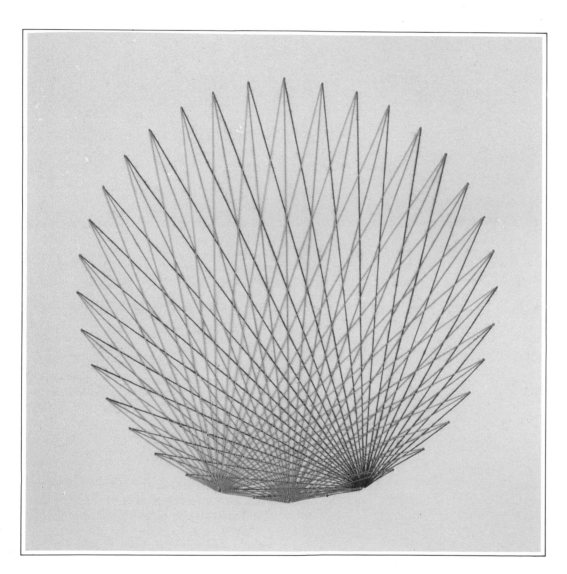

Mark out a 36 point circle.

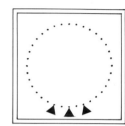

Mark the bottom point and the points two steps away on either side.

Sew each marked point to every other point.
The back of this design looks very much like the front.

Sunrise uses all the diagonals from three points grouped closely together at the bottom of the circle. The effect is very different and it has been enhanced by using three closely related colours.

Although incomplete mystic roses are not strictly curve stitching, in that a curve is not generated by sewing straight lines, the result created by following a simple rule can be very effective.

7. Sunflower

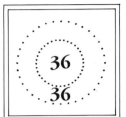

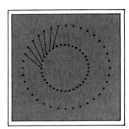

Mark out 2 concentric 36 point circles whose points lie on the same radial lines.

Sew from each point on the outer circle to a point 2 steps further round on the inner circle.

After completing the whole circle, continue by sewing again from each point on the outer circle to the point 6 steps further round in the same direction on the inner circle.

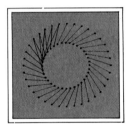

The sunflower is very simple to sew and it can easily be modified by changing the size of the inner circle and by altering the rules about the number of steps further round. It can also be sewn to create clockwise versions.

This pattern is most attractive, but it is not strictly curve stitching as both circles arise only from the initial drawing and not from the threads. However, the petals only appear as the design is sewn.

8. Two linked circles

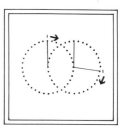

Mark out 2 equally sized 36 point circles that interlock for a third of their circumferences. (12 steps)

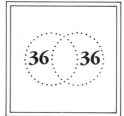

Number each circle clockwise, with the "1" starting ten steps further round on the right hand circle.

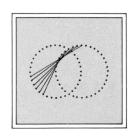

Sew from 1 to 1, 2 to 2, up to 36 to 36.

This pattern is just one representative of a whole family of possible curves which can be generated from two linked circles using essentially the same approach. With the circular template, it is easy to adjust the drawing so that any number of points interlock between the circles. The other source of variation is where to start numbering.

On page 38, there is a further development of this idea using three circles.

9. Concentric circles 1

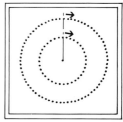

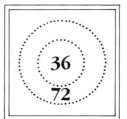

Mark out a 72 point circle with a diameter that is twice that of an inner 36 point circle. Every second point on the outer circle should lie on the same radial line as a point on the inner circle.

Number each circle clockwise starting at the same radius, the outer from 1 to 72 and the inner from 1 to 36.

Sew from 1 to 1, 2 to 2, up to 36 to 36. Continue by going round the inner circle for a second time joining 1 to 37, 2 to 38 and so on, finishing with 36 to 72.

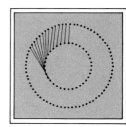

Complicated as this pattern appears to be, the back is extremely simple. The only stitches showing are between alternate pairs of adjacent points on each circumference. The centre portion rather resembles a cardioid (see page 44), but is not defined by the same sewing rule and so the resemblance is only superficial.

10. Concentric circles 2

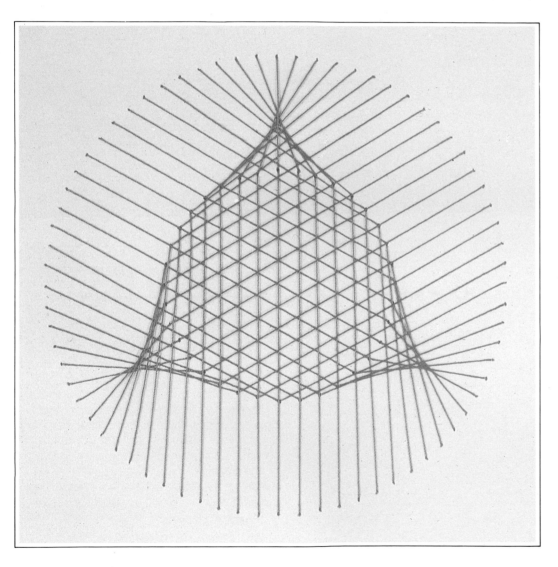

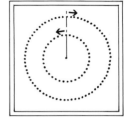

Mark out a 72 point circle with a diameter that is twice that of an inner 36 point circle. Every second point on the outer circle should lie on the same radial line as a point on the inner circle.

Starting at the same radius, number the outer circle clockwise from 1 to 72 and the inner circle anticlockwise from 1 to 36.

Sew from 1 to 1, 2 to 2, finishing with 36 to 36. Continue by going round the inner circle for a second time joining 1 to 37, 2 to 38, finishing with 36 to 72.

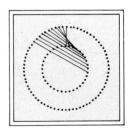

Although the sewing rule here is exactly the same as for the pattern overleaf, the finished result is remarkably different. The only change was to number the inner circle anticlockwise rather than clockwise. The stitches showing on the back are also between alternate pairs of adjacent points.

4.
Further developments

Further developments

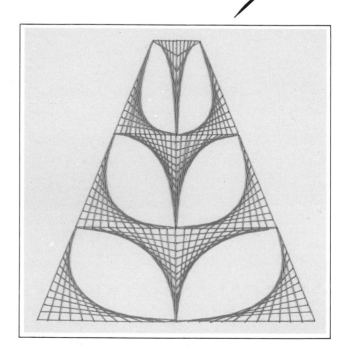

The interesting problem of searching for ways to sew curves of special mathematical significance is explored in this section. This means finding some property of each curve which generates an envelope without too much drawing and measuring. The most satisfying designs to sew are those where an important mathematical property of the curve is illustrated by following a simple rule.

This section contains a collection of more advanced patterns each complete in itself, but also chosen to suggest interesting lines of approach and experimentation. For instance, the linked parabola designs on this page could be copied as they stand, but it is far more valuable to use them to stimulate your own ideas. The concept of the hidden outline, illustrated by the design above, is an idea with much potential for further development.

1. Equal chords in a square

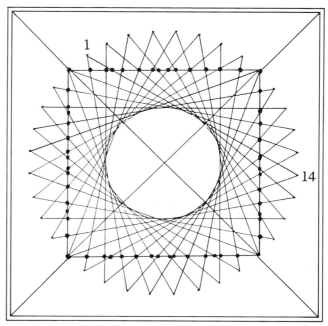

This pattern is very impressive because it seems remarkable that threads taken from what appear to be random points on the sides of a square none the less give a perfect circle at the centre.

The diagram of the back of the card solves the mystery. Draw an equal chord design in a circle and then superimpose a square on it. Carefully mark the points where the chords cross the square and prick holes at those points. Then sew only those portions of the chords which come within the square. Here a 36 point circle has been used together with a 13 step equal chord pattern. The number of points must be a multiple of 4.

An easy adaptation of this design would be to start with a 36 point outer circle and then to use the outline of a hexagon rather than a square, or a 40 point circle and an octagon.

2. Three linked circles

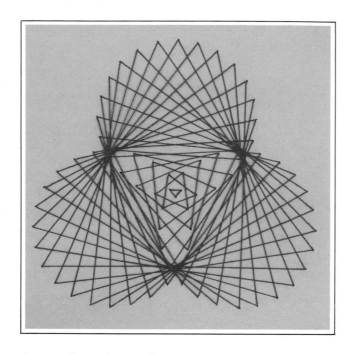

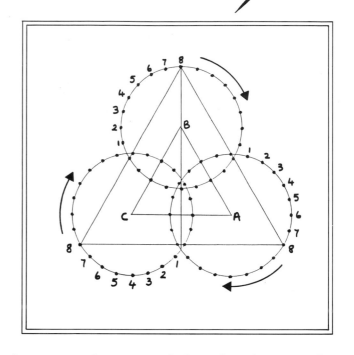

Draw the equilateral triangle A B C, and then construct three circles centred on its corners each with radius ⅗ the length of the sides of the triangle. Now draw in the larger equilateral triangle which will help you to mark each circle with 24 points. Number each circle clockwise as shown.

Work on two circles at a time linking them by sewing from 1 to 1, 2 to 2 ... 24 to 24. First link circles A and B, then B and C and finally C and A. This is a development which follows on from two linked circles (page 32).

To obtain exactly this design you must follow these instructions precisely. However, there are many other ways of linking three circles and sewing points on the circumferences which lead to attractive results. Investigations of this kind are part of the delight of more advanced curve stitching.

3. Ellipse from inverses

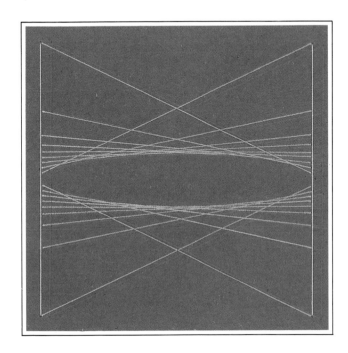

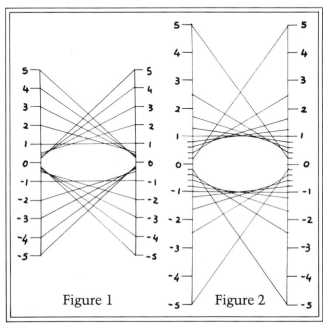

Figure 1   Figure 2

To construct an ellipse using the principle of inverses draw two parallel lines and join each number to its inverse on the opposite parallel, so 5 is joined to ⅕, 4 to ¼, 3 to ⅓, 2 to ½ and 1 to 1. Do the same with the negative numbers. This seems the obvious way to proceed and is illustrated in figure 1. However, it brings the points very close together as you approach zero and the threads tend to bunch up.

Figure 2 shows a rather better method. Divide the distance from −1 to 1 into equal parts (in this case 10) and sew across to their inverses on the opposite parallel. Thus ⅕ is joined to 5, ⅖ to 5⁄2, ⅗ to 5⁄3, ⅘ to 5⁄4 and 1 to 1. This gives a more even distribution of threads and is the one used for the pattern in the photograph.

Changing the distance between the parallel lines and the spacing of the numbering gives ellipses of varying eccentricity.

4. Ellipse from chords

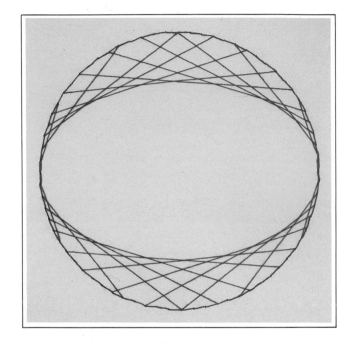

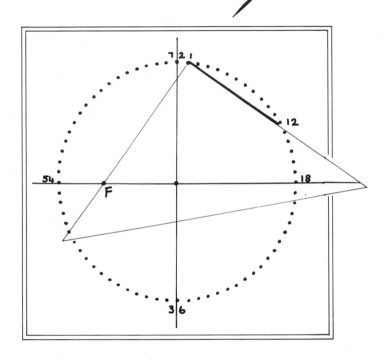

Another method of sewing the envelope of an ellipse is to mark out and number a 72 point circle. Choose a point F inside but not at the centre of the circle. In this example F lies on the horizontal diameter.

Place a set square on the 72 point circle so that its right angle is 'on a point on the circumference and F is on an adjacent side. Draw a chord from this point on the right angle as shown in the drawing. In theory all you do is repeat this process around the circle drawing chords which start on each of the 72 points.

However there are two snags about using this method when it comes to sewing. One is that it causes bunching of the chords towards points 18 and 54 where the chords have zero length. The other is that chords start but do not necessarily end on marked points. Both these problems can be avoided by dividing the circle into quadrants and then selecting only those points where the chord ends on or near another point. Once a chord like this has been found it can be repeated in the other three quadrants. You will find sufficient points in this way to generate an attractive ellipse.

In this illustration the original circle has also been stitched.

5. Hyperbola

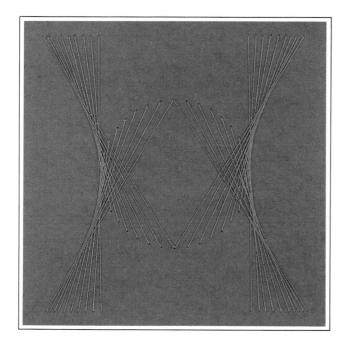

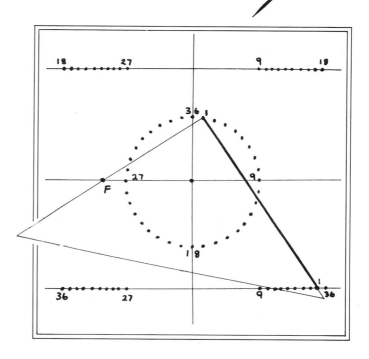

Draw 3 equally spaced parallel lines. Mark out and number a 36 point circle on the centre line whose circumference does not cut either of the other two lines. Draw a perpendicular line through the centre of the circle and then choose a point F which is on the horizontal centre line outside the circle.

As with the ellipse opposite place a set square on the 36 point circle so that its right angle is on a point on the opposite side of the circle from F. F lies on a side of the set square adjacent to the right angle. Draw a line from the point on the right angle to the furthest parallel line as shown in the diagram.

Working on one quadrant fill in a quarter of the pattern with a set square. Then copy those measurements with a pair of compasses on to the other three quadrants.

Notice how the points 18 and 36 on the circle are used twice while 9 and 27 are not used at all. The lines drawn from 9 and 27 on the parallels are tangents so are sewn straight from the top to the bottom line.

6. Curve of pursuit

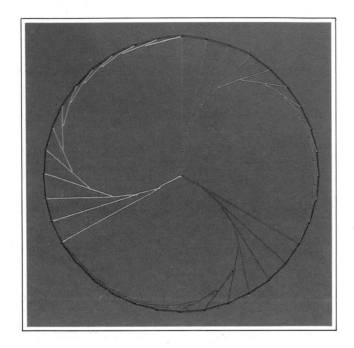

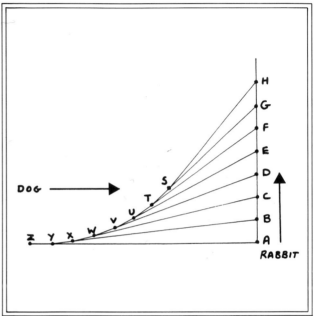

The classic way of describing a curve of pursuit is to imagine that a dog sees a rabbit crossing its path. He decides to give chase but, after running for a short distance, notices that the rabbit is no longer directly in front of him, so he changes direction to run towards where the rabbit now is. But the rabbit keeps running so the dog again has to change direction. This sequence will continue indefinitely if, as is assumed, they are both running at the same speed.

In the diagram the distances ZY, AB, YX, BC etc., are all the same length.

The photograph shows a pattern made from three identical curves of pursuit sewn within a 36 point circle. Designs like this can be further developed with clockwise and anticlockwise versions being combined in various ways.

7. Tractrix

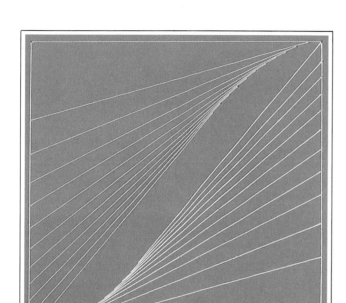

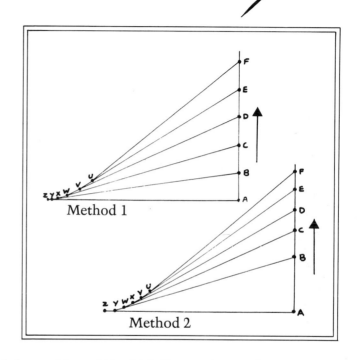

Method 1

Method 2

A tractrix, or tractory, is so called because it is the curve traced out by a rod being pulled by a person travelling in a straight line, with the other end of the rod not being on this line.

Although the tractrix resembles the curve of pursuit, it is harder to construct because each stitch must be the same length. The stitches represent the position of the rod after various intervals of time.

Using method 1, the end of the rod which is moving along the straight line is assumed to move at a constant speed and thus the lengths AB, BC, CD, DE etc. are all equal. Set a pair of compasses to the length of the rod AZ. With the compasses

find the position of Y, which lies on AZ, so that BY = AZ. Then find X, which lies on BY, so that CX = BY. At each stage the new position of the end of the rod lies on the previous position of the rod.

Method 2 shows what happens if instead the distances at the other end of the rod are supposed equal. The lengths ZY, YX, XW, WV etc. are now equal and the compasses are used to find the positions of A, B, C, D, E etc. This second method has been used twice to make the design in the photograph. Two tractrix curves have been sewn within a square, leaving an intriguing shape along a diagonal.

8. Cardioid

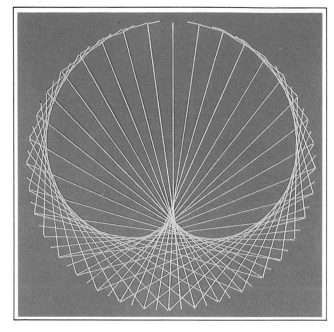

1 – 2	25 – 50	49 – 26
2 – 4	26 – 52	50 – 28
3 – 6	27 – 54	51 – 30
4 – 8	28 – 56	52 – 32
5 – 10	29 – 58	53 – 34
6 – 12	30 – 60	54 – 36
7 – 14	31 – 62	55 – 38
8 – 16	32 – 64	56 – 40
9 – 18	33 – 66	57 – 42
10 – 20	34 – 68	58 – 44
11 – 22	35 – 70	59 – 46
12 – 24	36 – 72	60 – 48
13 – 26	37 – 2	61 – 50
14 – 28	38 – 4	62 – 52
15 – 30	39 – 6	63 – 54
16 – 32	40 – 8	64 – 56
17 – 34	41 – 10	65 – 58
18 – 36	42 – 12	66 – 60
19 – 38	43 – 14	67 – 62
20 – 40	44 – 16	68 – 64
21 – 42	45 – 18	69 – 66
22 – 44	46 – 20	70 – 68
23 – 46	47 – 22	71 – 70
24 – 48	48 – 24	

The cardioid is the simplest of the family of curves called epicycloids. It has a single cusp and derives its name from its heart-like shape.

An epicycloid is the path traced out by a point on the circumference of a circle as it rolls around the outside of a fixed circle. The curve obtained is a cardioid when the fixed circle and the moving circle have the same radius.

This design creates a cardioid by generating the envelope within a larger circle which has a radius that is three times that of the moving circle.

Having marked out a 72 point circle numbered so that 72 is at the top, sew each number to its double. This is straightforward with stitches as far as 36 – 72. The next stitch 37 – 74 is sewn from 37 – 2 because 74 = 72+2. The table above shows all the joinings including the repeated chord 48 – 24, which has been sewn only once. It will save thread and keep the back neat if you use the nearest available point to start your next stitch. Cross off the stitches from the table as you sew them.

9. Nephroid

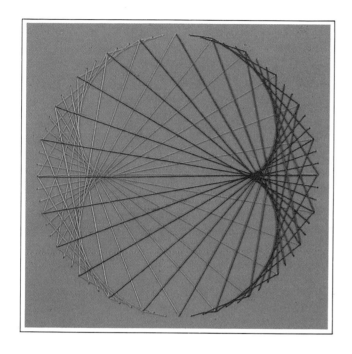

1 – 3	25 – 3	49 – 3
2 – 6	26 – 6	50 – 6
3 – 9	27 – 9	51 – 9
4 – 12	28 – 12	52 – 12
5 – 15	29 – 15	53 – 15
6 – 18	30 – 18	54 – 18
7 – 21	31 – 21	55 – 21
8 – 24	32 – 24	56 – 24
9 – 27	33 – 27	57 – 27
10 – 30	34 – 30	58 – 30
11 – 33	35 – 33	59 – 33
12 – 36	Change Colour	60 – 36
13 – 39	37 – 39	61 – 39
14 – 42	38 – 42	62 – 42
15 – 45	39 – 45	63 – 45
16 – 48	40 – 48	64 – 48
17 – 51	41 – 51	65 – 51
18 – 54	42 – 54	66 – 54
19 – 57	43 – 57	67 – 57
20 – 60	44 – 60	68 – 60
21 – 63	45 – 63	69 – 63
22 – 66	46 – 66	70 – 66
23 – 69	47 – 69	71 – 69
24 – 72	48 – 72	

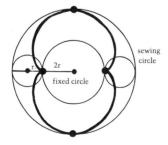

A nephroid is an epicycloid with two cusps. It is so called because of its kidney-like shape and is generated by a point on a circle rolling round a fixed circle twice its radius.

This design creates a nephroid by generating the envelope within a larger circle which has a radius that is four times that of the moving circle.

Having marked out a 72 point circle numbered so that 72 is at the top, sew each number to its treble. This is straightforward with stitches as far as 24-72. The next stitch 25-75 is sewn from 25-3, because 75 = 72+3. The table above shows all the joinings. There is no need to sew the stitches in the order of the table. It will save thread and keep the back neat if you use the nearest available point to start your next stitch. Cross off the stitches from the table as you sew them.

Change colour half way through sewing this pattern and repeat the chord 18-54 which occurs again in the second half. The horizontal axis is the only one which is sewn with two threads.

10. Epicycloid of Cremona

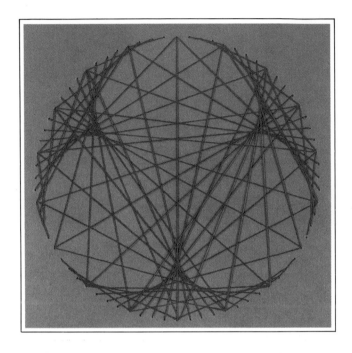

1 – 4	19 – 4	37 – 4	55 – 4
2 – 8	20 – 8	38 – 8	56 – 8
3 – 12	21 – 12	39 – 12	57 – 12
4 – 16	22 – 16	40 – 16	58 – 16
5 – 20	23 – 20	41 – 20	59 – 20
6 – 24	24 – 24*	42 – 24	60 – 24
7 – 28	25 – 28	43 – 28	61 – 28
8 – 32	26 – 32	44 – 32	62 – 32
9 – 36	27 – 36	45 – 36	63 – 36
10 – 40	28 – 40	46 – 40	64 – 40
11 – 44	29 – 44	47 – 44	65 – 44
12 – 48	30 – 48	48 – 48*	66 – 48
13 – 52	31 – 52	49 – 52	67 – 52
14 – 56	32 – 56	50 – 56	68 – 56
15 – 60	33 – 60	51 – 60	69 – 60
16 – 64	34 – 64	52 – 64	70 – 64
17 – 68	35 – 68	53 – 68	71 – 68
18 – 72	36 – 72	54 – 72	72 – 72*

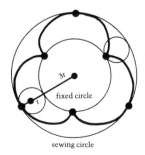

The epicycloid of Cremona, which is called after a mathematician of that name, has three cusps and is generated by a point on a circle rolling round a fixed circle three times its radius.

This design creates a three-cusped epicycloid by generating the envelope within a larger circle which has a radius that is five times that of the moving circle.

Mark out a 72 point circle numbered so that 72 is at the top and then consult the table above. Each point is joined to its quadruple. The calculation of quadruples is straightforward as far as 18–72, but 19–76 becomes 19–4 (because 76 = 72+4), 37–148 becomes 37–4 (because 148 = 72+72+4) and 55–220 becomes 55–4 (because 220 = 72+72+72+4). The table above gives the complete set of chords. The chords marked * do not have to be sewn as their lengths are zero. Do not try to sew in order, but go to the nearest available point. This will save thread and keep the back neat. Cross off each chord as you sew it.

Note that the cardioid, which was obtained by sewing each number to its double had one cusp. The nephroid which used the treble had two cusps and the epicycloid of Cremona which used the quadruple had three cusps.

11. Cycloid

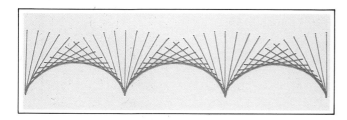

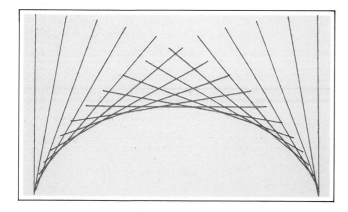

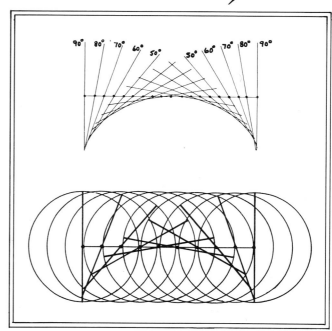

A cycloid is the path traced out by a point on the circumference of a circle as it rolls along a straight line. From a curve stitching point of view it is more interesting to sew the half height cycloid which is the envelope of the diameter. Each stitch is the diameter of the circle as it rolls along the straight line. To find the correct sequence of positions, we have to use the property that the centre of the circle also moves on a straight line and that equal rotations of the circle give equal movements of the centre along this line. If the rotations are to be 10° each, then the centre line should be divided into 18

equal divisions (180° ÷ 18 = 10°). The angles of the diameters are then 90°, 80°, 70°, 60°,..., 10°, 0°, 10°,..., 60°, 70°, 80°, 90°. The length of each stitch is related to the length of the centre line. If the length of the centre line is r, then each stitch is $2r \div \pi$. (If r = 90mm, then each stitch is 57mm long).

Having sewn one cycloid successfully it is then easy to develop the idea further and to generate more complex patterns linking several arches into a single design.

12. Deltoid

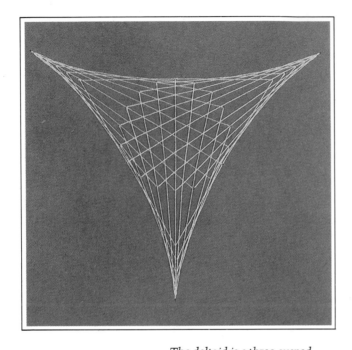

	FIRST CUSP		SECOND CUSP	THIRD CUSP
A	1 – 36	Tangent	12 – 12 T	24 – 24 T
B	1 – 34		13 – 10	25 – 22
C	2 – 32		14 – 8	26 – 20
D	3 – 30		15 – 6	27 – 18
E	4 – 28		16 – 4	28 – 16
F	5 – 26		17 – 2	29 – 14
G	6 – 24	Diameter	18 – 36 D	30 – 12 D
F	7 – 22		19 – 34	31 – 10
E	8 – 20		20 – 32	32 – 8
D	9 – 18		21 – 30	33 – 6
C	10 – 16		22 – 28	34 – 4
B	11 – 14		23 – 26	35 – 2
A	12 – 12	Tangent	24 – 24 T	36 – 36 T

Mark out a 36 point circle and number it so that 36 is at the top. The rule for calculating the table above is to double each number and then subtract it from 36 (the number of points on the circle). Subtract numbers larger than 36 from 72. The chords in the table do not generate the envelope of the deltoid within the circle, but outside it, so this design must be partially constructed by drawing on the back of the card before starting to sew.

Draw the first cusp by following the first column in the table.

1. Draw the diameter 6–24 and extend it beyond 6 to double its length, calling the end point G. Draw tangents at points 36 and 12, calling the point where they meet on the extended diameter A.

2. Next draw the chords 1–34 and 11–14, extending them outside the circle. They meet on the extended diameter at B.

3. Find C, D, E and F by drawing the corresponding pairs of secants, which meet on the extended diameter.

4. Now draw in the other two extended diameters and use compasses to transfer the distances of the points A, B, C, D, E, F and G from the first cusp to the other two cusps.

The diagram opposite should help.

For each stitch, only the numbered point on the blue background is pricked and sewn through. The stitch passes over the other point and goes to the corresponding lettered point in the cusp. For instance, the stitch 2–32 is sewn through 32, but passes over 2 and is sewn through C. The stitch 3–30 is sewn through 30, but passes over 3 and is sewn through D. Complete one cusp at a time. For this design only the even numbered points are ever pricked through.

The deltoid is a three-cusped member of the family of curves called hypocycloids. A hypocycloid is the path traced out by a point on the circumference of a circle as it rolls round the inside of a fixed circle. The curve obtained is a deltoid when the fixed circle has a radius three times that of the moving circle.
This design creates a deltoid by generating the envelope outside a smaller circle which has a radius one third that of the fixed circle. The smaller circle has the same radius as the moving circle.

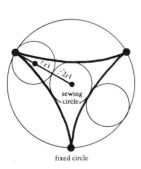

fixed circle

13. Four deltoids

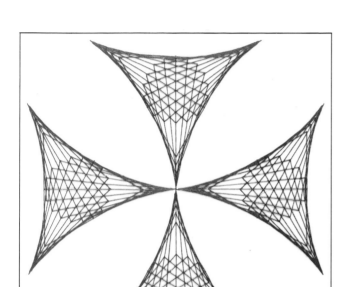

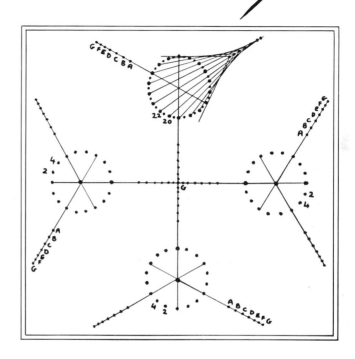

Once you have sewn one deltoid successfully, you will realise that it is far easier to do than to explain. This very attractive pattern is made by sewing four deltoids so that their cusps meet at the centre. Complete the design by following the instructions opposite.

Draw one 36 point circle, **number it** so that 36 is at the top and construct the measurements for the points A, B, C, D, E, F and G on one extended diameter. The other circles need only have 18 points, which should be **marked with** even numbers 2, 4, 6, etc. Use compasses to transfer the **distances** of the points A, B, C, D, E, F and G from the first extended **diameter** to the other eleven.

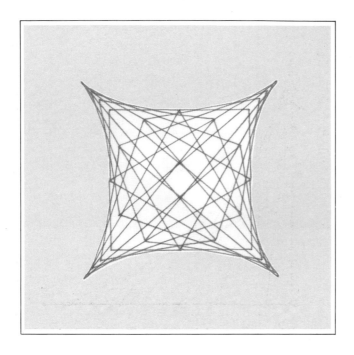

	FIRST CUSP	SECOND CUSP	THIRD CUSP	FOURTH CUSP
A	36 – 36 Tangent	9 – 9 T	18 – 18 T	27 – 27 T
B	1 – 33	10 – 6	19 – 15	28 – 24
C	2 – 30	11 – 3	20 – 12	29 – 21
D	3 – 27	12 – 36	21 – 9	30 – 18
E	4 – 24	13 – 33	22 – 6	31 – 15
	Diameter			
E	5 – 21	14 – 30	23 – 3	32 – 12
D	6 – 18	15 – 27	24 – 36	33 – 9
C	7 – 15	16 – 24	25 – 33	34 – 6
B	8 – 12	17 – 21	26 – 30	35 – 3
A	9 – 9 Tangent	18 – 18 T	27 – 27 T	36 – 36 T

Mark out a 36 point circle and number it so that 36 is at the top. The rule for calculating the table above is to treble each number and then to subtract it from 36 (the number of points on the circle). Subtract numbers larger than 36 from 72 or 108.

The chords in the table do not generate the envelope of the astroid within the circle but outside it, so this design must be partially constructed by drawing on the back of the card before starting to sew.

Although the diameters are not actually sewn, they are needed in the construction.

Draw the first cusp by following the first column in the table.

1. Draw the diameter 22½ – 4½ and extend it at 4½ beyond the circle half as far again. Draw the tangents at points 36 and 9, and call the point where they meet on the extended diameter A. Points B, C, D and E also lie on this extension.

2. Next draw the chords 1–33 and 8–12, extending them outside the circle. They meet on the extended diameter at B.

3. Find C, D and E by drawing the corresponding pairs of secants, which meet on the extended diameter.

4. Draw the other diameter 13½ – 31½ and extend it both ways outside the circle. Extend the first diameter outside the circle the other way. Use compasses to transfer the distances of the points A, B, C, D and E from the first cusp to the other three cusps.

For each stitch, only the numbered point on the blue background is pricked and sewn through. The stitch passes over the other point and goes on to the corresponding lettered point in the cusp. For instance 1 – 33 is sewn through 33, but passes over 1 and is sewn through B. The stitch 2 – 30 is sewn through 30, but passes over 2 and is sewn through C. Complete one cusp at a time. Only those points on the circle where the number divides by 3 are pricked and sewn through.

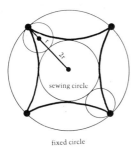

sewing circle

fixed circle

The astroid is a hypocycloid with four cusps and is generated by a point on a circle rolling round the inside of a fixed circle four times its radius.

This design creates an astroid by generating the envelope outside a smaller circle which has a radius one half that of the fixed circle. The smaller circle has twice the radius of the moving circle.

15. Astroid 2

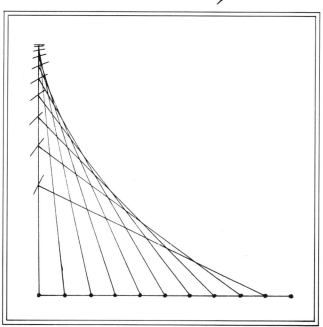

This is a completely different method of constructing an astroid. Think of it as a curve generated by a ladder sliding down a wall. The diagram on the right shows how to construct a quarter of the design using equally spaced points along the horizontal axis. The points on the vertical axis are found with compasses set to the length of the ladder.

At first glance it may seem as if the envelope is a parabola, but in fact the astroid is a different curve. The difference comes about because the divisions are not equally spaced on the vertical axis.

There is a similarity between the tractix (page 43) and this astroid because both have constant stitch length. However the astroid stitch always lies between the two axes while the tractrix stitch lies between the axis and the previous stitch.

The parabola, the curve of pursuit, the tractrix and the astroid are appear together on page 72.

16. Six-cusped hypocycloid

	FIRST CUSP		SECOND CUSP	THIRD CUSP
A	36 – 72	Tangent	12 – 12 T	24 – 24 T
B	1 – 67		13 – 7	25 – 19
C	2 – 62		14 – 2	26 – 14
D	3 – 57		15 – 69	27 – 9
E	4 – 52		16 – 64	28 – 4
F	5 – 47		17 – 59	29 – 71
G	6 – 42	Diameter	18 – 54 D	30 – 66 D
F	7 – 37		19 – 49	31 – 61
E	8 – 32		20 – 44	32 – 56
D	9 – 27		21 – 39	33 – 51
C	10 – 22		22 – 34	34 – 46
B	11 – 17		23 – 29	35 – 41
A	12 – 12	Tangent	24 – 24 T	36 – 36 T

This table produces three of the cusps which is half the design. It will probably be easy enough to sew the remainder without consulting the table.

Mark out a 72 point circle and number it so that 72 is at the top. The rule for calculating the table is to multiply each number by 5 and then subtract it from 72 (the number of points on the circle). Subtract numbers larger than 72 from 144, 216, 288 or 360.

The chords in the table do not generate the envelope of the hypocyloid within the circle, but outside it, so this design must be partially constructed on the back of the card before starting to sew.

Draw the first cusp by following the first column in the table.

1. Draw the diameter 6–42 and extend it at 6 beyond the circle a quarter as far again, calling the end point G. Draw tangents at points 72 and 12 and call the point where they meet on the extended diameter A. Points B, C, D, E and F also lie on this extension.

2. Next draw the chords 1–67 and 11–17, extending them outside the circle. They meet on the extended diameter at B.

3. Find C, D, E and F by drawing the corresponding pairs of secants which meet on the extended diameter. G is so close to F that they are treated as the same point in this design.

4. Extend the 6–42 diameter beyond 42, and now draw in the other two diameters and extend them both ways. Use compasses to transfer the distances A, B, C, D, E, F and G from the first cusp to the other five cusps.

For each stitch, only the numbered point on the blue background is pricked and sewn through. The stitch passes over the other point and goes to the corresponding lettered point in the cusp. For instance, the stitch 1–67 is sewn through 67, but passes over 1 and is sewn through B. The stitch 2 – 62 is sewn through 62, but passes over 2 and is sewn through C. Complete one cusp at a time. Because 72 is not a multiple of 5, all the points on the circle except those on the diameters are pricked and sewn, but each one with only one thread. It is interesting to investigate what happens if the original number of points is chosen to be 60, a multiple of both 5 and 6.

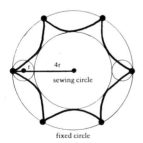

fixed circle

The six-cusped hypocycloid is the path traced out by a point on the circumference of a circle as it rolls around the inside of a fixed circle which has a radius six times that of a moving circle.

This design creates a six-cusped hypocycloid by generating the envelope outisde a smaller circle which has a radius two thirds that of the fixed circle. The smaller circle has four times the radius of the moving circle.

5.
Computer programs

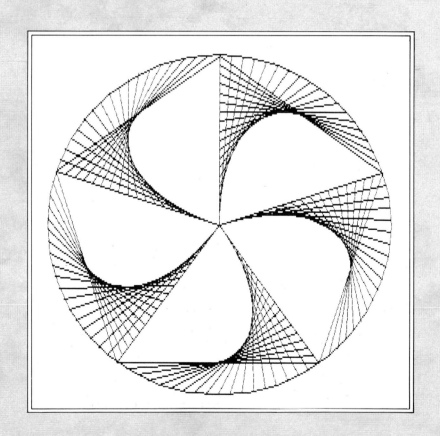

Programs

Notes on computer programs

Since these programs are very short, they will take you only a few minutes to type. Each can be viewed in its entirety on the screen which makes changes easier to carry out, as does the fact that no line overflows on to a second line.

Computers work in radians rather than degrees and the programs make extensive use of 2 π radians which is equivalent to 360°. At times slightly more or less than 2 π is needed which is why you might find 6.29 or 6.28 being used instead.

Numeric variables are assigned in the same way throughout most of the programs as follows:

Co-ordinates	A,D and X,Y
Groups of co-ordinates	J,K,L,M and P,Q,R,S
Circles	B
Cusps	C
Focus	F
Inputs	N,V
General purpose	G,H,T.

Inputs have been kept to a minimum but, if you want, they can be reduced further. For example, try replacing a line such as:
INPUT "How many points? "V by V=36 (for the BBC) or
INPUT "How many points? ";V by LET V=36 (for the Spectrum).

If you want to investigate the effect of altering a circle to an ellipse, this can easily be done by making some slight changes. For instance, you could have a mystic elliptical rose rather than a mystic rose by replacing the first 500 by, say, 630 in lines 60 and 70 of that program (for the BBC). For the Spectrum, replace 87 by 110 in line 60 and the first 87 by 110 in line 90.

End points of the chords for epicycloids and starting points of the secants for hypocycloids are displayed down the left-hand side of the screen. If you do not need these numbers then leave

out lines 120 to 140 of the BBC program concerned. For the Spectrum, leave out lines 70 to 120 of the epicycloid program or lines 90 to 140 of the hypocycloid program. To save space, the other point in each pair is not shown as these points run consecutively from 1 up to the number you entered. In the spiral program you can omit lines 150 and 160 of the BBC version to suppress the numbers or, for the Spectrum, lines 160 to 190.

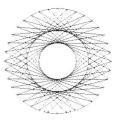 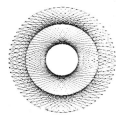

BBC Programs Only
In most of the programs requiring an input from you, the screen will clear after you have entered the appropriate numbers and then the computer will start drawing the pattern. Should you prefer your data to stay on the screen all the time, simply move the MODE 1 or 0 from wherever it is to line 5.

Many of the circle-based programs contain six or so lines near the beginning for drawing a circle in red round the design. They run from GCOL 0, 1 to GCOL 0, 3 inclusive and are optional.

Mode 1 gives good resolution on the screen and Mode 0 is even better, but for more colourful displays you need the lower resolution Mode 2. For this, use lines 20, 30 and 50 (without REPEAT) of the Astroid program after you have renumbered them to fit in with the rest of your program. Try moving the GCOL 0, RND(7) to a higher line number for some multicoloured effects.

Programs

---

1. Parabolas in a polygon

Choose the number of sides
Choose the number of steps per side

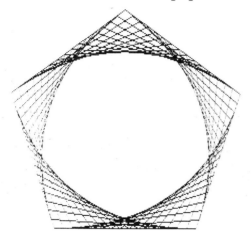

5 sides
15 steps per side

B.B.C.

```
  0 REM Parabolas - Polygon
 10 INPUT'"How many sides? "V
 20 INPUT'"And steps per side? "N
 30 MODE 1
 40 FOR B=0 TO 6.28 STEP 2*PI/V
 50 G=B+2*PI/V
 60 T=B+4*PI/V
 70 FOR H=500/N TO 501 STEP 500/N
 80 P=(500-H)*SIN B+H*SIN G+640
 90 Q=(500-H)*COS B+H*COS G+500
100 X=(500-H)*SIN G+H*SIN T+640
110 Y=(500-H)*COS G+H*COS T+500
120 MOVE P,Q
130 DRAW X,Y
140 NEXT H,B
```

Spectrum

```
 10 REM Parabolas - Polygon
 20 INPUT "How many sides? ";V
 30 INPUT "Steps per side? ";N
 40 FOR B=0 TO 6.28 STEP 2*PI/V
 50 LET G=B+2*PI/V
 60 LET T=B+4*PI/V
 70 FOR H=87/N TO 88 STEP 87/N
 80 LET K=87-H
 90 LET P=K*SIN B+H*SIN G+127
100 LET Q=K*COS B+H*COS G+87
110 LET X=K*SIN G+H*SIN T+127
120 LET Y=K*COS G+H*COS T+87
130 PLOT P,Q: DRAW X-P,Y-Q
140 NEXT H: NEXT B
```

2. Parabolas in six polygons

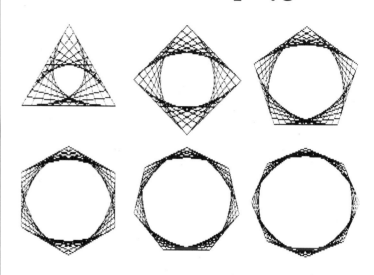

3–8 sides

B.B.C.

```
  0 REM Parabolas - 6 Polygons
 10 MODE 0
 20 V=3
 30 FOR D=750 TO 250 STEP -500
 40 FOR A=200 TO 1072 STEP 436
 50 FOR B=0 TO 6.28 STEP 2*PI/V
 60 G=B+2*PI/V
 70 T=B+4*PI/V
 80 FOR H=0 TO 180 STEP 20
 90 P=(200-H)*SIN B+H*SIN G+A
100 Q=(200-H)*COS B+H*COS G+D
110 X=(200-H)*SIN G+H*SIN T+A
120 Y=(200-H)*COS G+H*COS T+D
130 MOVE P,Q
140 DRAW X,Y
150 NEXT H,B
160 V=V+1
170 NEXT A,D
```

Spectrum

```
 10 REM Parabolas - 6 Polygons
 20 LET V=3
 30 FOR D=134 TO 41 STEP -93
 40 FOR A=40 TO 214 STEP 87
 50 FOR B=0 TO 6.28 STEP 2*PI/V
 60 LET G=B+2*PI/V
 70 LET T=B+4*PI/V
 80 FOR H=0 TO 35 STEP 5
 90 LET K=40-H
100 LET P=K*SIN B+H*SIN G+A
110 LET Q=K*COS B+H*COS G+D
120 LET X=K*SIN G+H*SIN T+A
130 LET Y=K*COS G+H*COS T+D
140 PLOT P,Q: DRAW X-P,Y-Q
150 NEXT H: NEXT B
160 LET V=V+1
170 NEXT A: NEXT D
```

Programs

3. Parabolic star

Choose the number of points
Choose the number of steps per line

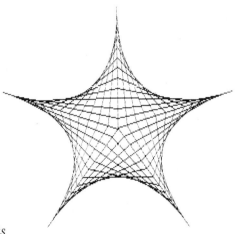

5 points
12 steps per line

B.B.C.

```
  0 REM Parabolas - Star
 10 INPUT'"How many points? "V
 20 INPUT'"And steps per line? "N
 30 MODE 1
 40 FOR B=0 TO 6.28 STEP 2*PI/V
 50 T=B+2*PI/V
 60 FOR H=500/N TO 501 STEP 500/N
 70 G=500-H
 80 MOVE G*SIN B+640,G*COS B+500
 90 DRAW H*SIN T+640,H*COS T+500
100 NEXT H,B
```

Spectrum

```
 10 REM Parabolas - Star
 20 INPUT "How many points? ";V
 30 INPUT "Steps per line? ";N
 40 FOR B=0 TO 6.28 STEP 2*PI/V
 50 LET T=B+2*PI/V
 60 FOR H=87/N TO 88 STEP 87/N
 70 LET X=(87-H)*SIN B
 80 LET Y=(87-H)*COS B
 90 PLOT X+127,Y+87
100 DRAW H*SIN T-X,H*COS T-Y
110 NEXT H: NEXT B
```

4. Six parabolic stars

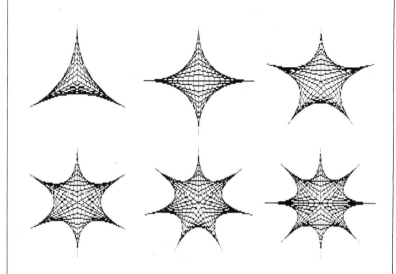

3–8 cusps

B.B.C.

```
  0 REM Parabolas - 6 Stars
 10 MODE 0
 20 V=3
 30 FOR D=750 TO 250 STEP -500
 40 FOR A=200 TO 1072 STEP 436
 50 FOR B=0 TO 6.28 STEP 2*PI/V
 60 T=B+2*PI/V
 70 FOR H=0 TO 180 STEP 20
 80 G=200-H
 90 MOVE G*SIN B+A,G*COS B+D
100 DRAW H*SIN T+A,H*COS T+D
110 NEXT H,B
120 V=V+1
130 NEXT A,D
```

Spectrum

```
 10 REM Parabolas - 6 Stars
 20 LET V=3
 30 FOR D=134 TO 41 STEP -93
 40 FOR A=40 TO 214 STEP 87
 50 FOR B=0 TO 6.28 STEP 2*PI/V
 60 LET T=B+2*PI/V
 70 FOR H=0 TO 35 STEP 5
 80 LET X=(40-H)*SIN B
 90 LET Y=(40-H)*COS B
100 PLOT X+A,Y+D
110 DRAW H*SIN T-X,H*COS T-Y
120 NEXT H: NEXT B
130 LET V=V+1
140 NEXT A: NEXT D
```

57

Programs

5. Curves in a circle

Choose the number of sectors
Choose the number of points per sector

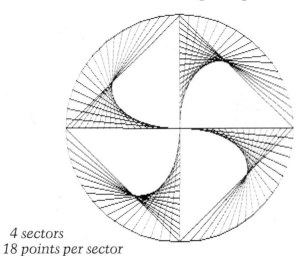

4 sectors
18 points per sector

B.B.C.

```
  0 REM Curves - Circle
 10 INPUT'"How many sectors? "V
 20 INPUT'"And points per sector? "N
 30 MODE 1
 40 GCOL 0,1
 50 MOVE 640,1000
 60 FOR B=0 TO 6.29 STEP PI/18
 70 DRAW 500*SIN B+640,500*COS B+500
 80 NEXT B
 90 GCOL 0,3
100 K=V*N
110 FOR B=0 TO 6.28 STEP 6.29/V
120 FOR H=B TO 6.29/V+B STEP 2*PI/K
130 T=250*(H-B)*V/PI
140 MOVE T*SIN B+640,T*COS B+500
150 DRAW 500*SIN H+640,500*COS H+500
160 NEXT H,B
```

Spectrum

```
 10 REM Curves - Circle
 20 INPUT "How many sectors?";V
 30 INPUT "Points per sector";N
 40 CIRCLE INK 2,127,87,87
 50 LET K=2*PI/(V*N)
 60 FOR B=0 TO 6.28 STEP 6.29/V
 70 FOR H=B TO 6.29/V+B STEP K
 80 LET T=43*(H-B)*V/PI
 90 LET X=T*SIN B
100 LET Y=T*COS B
110 PLOT X+127,Y+87
120 DRAW 86*SIN H-X,86*COS H-Y
130 NEXT H: NEXT B
```

6. Curves in six circles

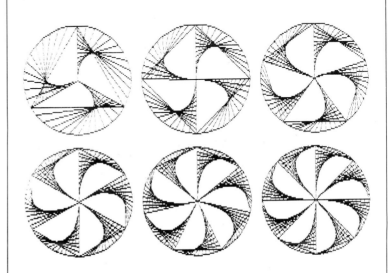

3–8 sectors

B.B.C.

```
  0 REM Curves - 6 Circles
 10 MODE 0
 20 V=3
 30 FOR D=750 TO 250 STEP -500
 40 FOR A=200 TO 1072 STEP 436
 50 MOVE A,D+200
 60 FOR B=0 TO 6.29 STEP PI/18
 70 DRAW 200*SIN B+A,200*COS B+D
 80 NEXT B
 90 FOR B=0 TO 6.28 STEP 6.29/V
100 FOR H=B TO 6.29/V+B STEP PI/(6*V)
110 T=100*(H-B)*V/PI
120 MOVE T*SIN B+A,T*COS B+D
130 DRAW 200*SIN H+A,200*COS H+D
140 NEXT H,B
150 V=V+1
160 NEXT A,D
```

Spectrum

```
 10 REM Curves - 6 Circles
 20 LET V=3
 30 FOR D=134 TO 41 STEP -93
 40 FOR A=40 TO 214 STEP 87
 50 CIRCLE A,D,41
 60 LET K=PI/(6*V)
 70 FOR B=0 TO 6.28 STEP 6.29/V
 80 FOR H=B TO 6.29/V+B STEP K
 90 LET T=20*(H-B)*V/PI
100 LET X=T*SIN B
110 LET Y=T*COS B
120 PLOT X+A,Y+D
130 DRAW 40*SIN H-X,40*COS H-Y
140 NEXT H: NEXT B
150 LET V=V+1
160 NEXT A: NEXT D
```

Programs

7. Eight parabolas in a square

Choose the number of steps

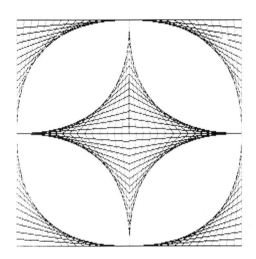

15 steps

B.B.C.

```
  0 REM 8 Parabolas
 10 INPUT'"How many steps/line? "V
 20 MODE 1
 30 N=V*(500 DIV V)
 40 FOR H=0 TO N STEP 500 DIV V
 50 FOR T=-1 TO 1 STEP 2
 60 FOR G=-1 TO 1 STEP 2
 70 FOR R=0 TO 1
 80 MOVE 640+H*T,N*(1+G*R)
 90 DRAW 640+N*T*R,N*(1+G)-H*G
100 NEXT R,G,T,H
```

Spectrum

```
 10 REM 8 Parabolas
 20 INPUT "Steps per line? ";V
 30 FOR H=0 TO 88 STEP 87/V
 40 FOR T=-1 TO 1 STEP 2
 50 FOR G=-1 TO 1 STEP 2
 60 FOR R=0 TO 1
 70 PLOT H*T+127,87*(1+G*R)
 80 LET K=87*T*R-H*T
 90 DRAW K,87*G*(1-R)-H*G
100 NEXT R: NEXT G
110 NEXT T: NEXT H
```

8. Parabola design

Choose random or your own numbers

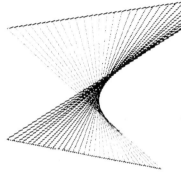

Spectrum

```
 10 REM Parabola - Design
 20 INPUT "R-Rnd or I-Input";T$
 30 LET J=INT (RND*255)
 40 LET L=INT (RND*255)
 50 LET P=INT (RND*255)
 60 LET R=INT (RND*255)
 70 LET K=INT (RND*160)
 80 LET M=INT (RND*160)
 90 LET Q=INT (RND*160)
100 LET S=INT (RND*160)
110 LET G=0: LET H=1: LET V=20
120 IF T$="R" THEN GO TO 180
130 INPUT "Start L.1 (x,y)";J,K
140 INPUT "End  L.1 (x,y)";L,M
150 INPUT "Start L.2 (x,y)";P,Q
160 INPUT "End  L.2 (x,y)";R,S
170 INPUT "Steps per line? ";V
180 CLS : LET G=1-G: LET H=1-H
190 PRINT J;",";K;"-";L;",";M;
200 PRINT "  ";P;",";Q;"-";R;
210 PRINT ",";S'" N-New & any";
220 PRINT " other to reverse"
230 PLOT J,K: DRAW L-J,M-K
240 PLOT P,Q: DRAW R-P,S-Q
250 FOR N=0 TO V
260 LET A=L-N*(L-J)/V
270 LET D=M-N*(M-K)/V
280 PLOT A,D
290 LET X=N*(R-P)/V
300 LET Y=N*(S-Q)/V
310 IF G THEN DRAW R-X-A,S-Y-D
320 IF H THEN DRAW P+X-A,Q+Y-D
330 NEXT N
340 PAUSE 0
350 IF INKEY$="N" THEN GO TO 30
360 GO TO 180
```

B.B.C.

```
  0 REM Parabola - Design
 10 INPUT'"R-Random or I-Input? "T$
 20 J=RND(1276):K=RND(960)
 30 L=RND(1276):M=RND(960)
 40 P=RND(1276):Q=RND(960)
 50 R=RND(1276):S=RND(960)
 60 G=0:H=1:V=30:IF T$="R" THEN 120
 70 INPUT'"Start of Line 1 (x,y)?"J,K
 80 INPUT'"End of Line 1 (x,y)?"L,M
 90 INPUT'"Start of Line 2 (x,y)?"P,Q
100 INPUT'"End of Line 2 (x,y)?"R,S
110 INPUT'"How many steps/line? "V
120 MODE 0:G=1-G:H=1-H
130 PRINT;J","K"-"L","M"  "P","Q"-"R;
140 PRINT;","S"  Press N for New & ";
150 PRINT "any other key to reverse";
160 MOVE J,K:DRAW L,M
170 MOVE P,Q:DRAW R,S
180 FOR N=0 TO V
190 MOVE L-N*(L-J)/V,M-N*(M-K)/V
200 IF G DRAW R-N*(R-P)/V,S-N*(S-Q)/V
210 IF H DRAW P+N*(R-P)/V,Q+N*(S-Q)/V
220 NEXT N
230 IF GET$="N" THEN 20 ELSE 120
```

59

Programs

9. Ellipse from inverses

Choose the width
Choose the height

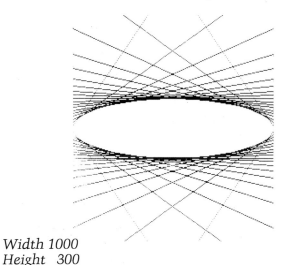

Width 1000
Height 300

B.B.C.

```
  0 REM Ellipse - Inverses
 10 INPUT'"Enter width (100-1276) "V
 20 INPUT'"And height (100-1016) "N
 30 MODE 1
 40 FOR T=-V/2 TO V/2 STEP V
 50 MOVE 640+T,0
 60 DRAW 640+T,1016
 70 FOR H=-10 TO 10
 80 MOVE 640+T,508+N*H/20
 90 IF H DRAW 640-T,508+5*N/H
100 NEXT H,T
```

Spectrum

```
 10 REM Ellipse - Inverses
 20 INPUT "Width? (255 max) ";V
 30 INPUT "Height?(174 max) ";N
 40 FOR T=-V/2 TO V/2 STEP V
 50 PLOT T+127,0: DRAW 0,174
 60 FOR H=-10 TO 10
 70 IF H=0 THEN GO TO 140
 80 LET R=N*H/20: LET K=5*N/H
 90 PLOT T+127, R+87: LET G=1
100 IF ABS K<=87 THEN GO TO 130
110 LET G=(87*SGN K-R)/(K-R)
120 LET K=87*SGN K
130 DRAW -2*T*G,K-R
140 NEXT H: NEXT T
```

10. Ellipse from chords

Choose the focus

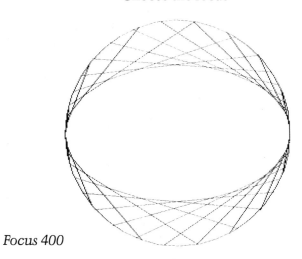

Focus 400

B.B.C.

```
  0 REM Ellipse - Chords
 10 INPUT'"Enter focus (150-450) "F
 20 MODE 1
 30 GCOL 0,1
 40 MOVE 628,1020
 50 FOR B=0 TO 6.29 STEP PI/18
 60 DRAW 500*SIN B+628,500*COS B+520
 70 NEXT B
 80 GCOL 0,3
 90 PRINT TAB(20-F/32,15)"."
100 G=0
110 FOR B=0 TO PI/4 STEP PI/120
120 K=31*PI/120-B
130 IF G<PI THEN G=G+B ELSE G=G+K
140 H=SQR(500^2+F^2-1000*F*COS G)
150 T=G+2*ASN(F*SIN G/H)
160 MOVE 500*COS G+628,500*SIN G+520
170 DRAW 500*COS T+628,500*SIN T+520
180 NEXT B
```

Spectrum

```
 10 REM Ellipse - Chords
 20 INPUT "Focus? (30-85) ";F
 30 CIRCLE INK 2,127,87,87
 40 PLOT 127-F,86
 50 PLOT 127-F,87
 60 LET G=0
 70 FOR B=0 TO PI/4 STEP PI/120
 80 IF G<=PI THEN LET G=G+B
 90 IF G>PI THEN LET G=G+PI/4-B
100 LET H=87^2+F^2-174*F*COS G
110 LET R=SQR H
120 LET T=G+2*ASN (F*SIN G/R)
130 LET X=87*COS G
140 LET Y=87*SIN G
150 PLOT X+127,Y+87
160 DRAW 87*COS T-X,87*SIN T-Y
170 NEXT B
```

Programs

11. Ellipse from circles

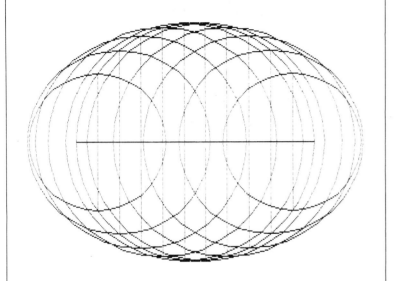

B.B.C.

```
  0 REM Ellipse - Circles
 10 MODE 1
 20 GCOL 0,1
 30 MOVE 1080,500
 40 FOR B=-PI/2 TO 3*PI/2 STEP PI/18
 50 DRAW 440*SIN B+640,440*COS B+500
 60 NEXT B
 70 DRAW 200,500
 80 GCOL 0,3
 90 FOR A=80 TO 800 STEP 80
100 H=SQR(880*A-A*A)
110 MOVE A+200,500-H
120 DRAW A+200,500+H
130 FOR B=0 TO 6.29 STEP PI/18
140 DRAW H*SIN B+A+200,H*COS B+500
150 NEXT B,A
```

Spectrum

```
 10 REM Ellipse - Circles
 20 CIRCLE INK 2,127,87,87
 30 PLOT 40,87
 40 DRAW 174,0
 50 FOR A=15 TO 159 STEP 16
 60 LET H=SQR (174*A-A*A)
 70 PLOT A+40,87-H
 80 DRAW 0,2*H
 90 CIRCLE A+40,87,H
100 NEXT A
```

12. Hyperbola

Choose the number of points
Choose the focus

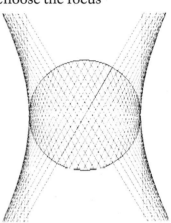

72 points
Focus 600

B.B.C.

```
  0 REM Hyperbola
 10 INPUT'"How many points? (Even) "V
 20 INPUT'"Enter focus (340-640) "F
 30 MODE 1
 40 GCOL 0,1
 50 MOVE 640,820
 60 FOR B=0 TO 6.29 STEP PI/18
 70 DRAW 300*SIN B+640,300*COS B+520
 80 NEXT B
 90 GCOL 0,3
100 PRINT TAB(20-F/32,15)"."
110 H=ASN(300/F)
120 G=2*(PI+2*H)/(V-2)
130 FOR B=-H TO H+3.15 STEP G
140 X=300*SIN B
150 Y=300*COS B
160 T=900*SGN Y
170 MOVE 640+X,520+Y
180 DRAW 640+X+T*Y/(F+X),520+Y-T
190 MOVE 640-X,520-Y
200 DRAW 640-X-T*Y/(F+X),520-Y+T
210 NEXT B
```

Spectrum

```
 10 REM Hyperbola
 20 INPUT "Points? (Even) ";V
 30 INPUT "Focus? (70-125) ";F
 40 CIRCLE INK 2,127,87,56
 50 PLOT 127-F,86
 60 PLOT 127-F,87
 70 LET H=ASN (55/F)
 80 LET G=2*(PI+2*H)/(V-2)
 90 FOR B=-H TO H+3.15 STEP G
100 LET X=55*SIN B
110 LET Y=55*COS B
120 LET T=Y+87*SGN Y
130 PLOT 127+X,87+Y
140 DRAW T*Y/(F+X),-T
150 PLOT 127-X,87-Y
160 DRAW -T*Y/(F+X),T
170 NEXT B
```

Programs

---

13. Curve of pursuit

Choose the length of step

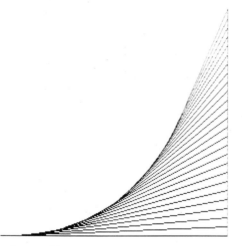

Length of step 40

B.B.C.

```
  0 REM Curve of Pursuit
 10 INPUT'"Length of step? (4-200) "L
 20 MODE 1
 30 X=0
 40 Y=0
 50 FOR N=0 TO 1000 STEP L
 60 MOVE X+100,Y
 70 DRAW 1200,N
 80 H=SQR((1100-X)^2+(N-Y)^2)
 90 X=X+L*(1100-X)/H
100 Y=Y+L*(N-Y)/H
110 NEXT N
120 DRAW 1200,0
```

Spectrum

```
 10 REM Curve of Pursuit
 20 INPUT "Step length(5-50)";L
 30 LET X=0
 40 LET Y=0
 50 FOR N=0 TO 175 STEP L
 60 LET P=200-X
 70 LET Q=N-Y
 80 PLOT X+25,Y
 90 DRAW P,Q
100 LET H=SQR (P^2+Q^2)
110 LET X=X+L*P/H
120 LET Y=Y+L*Q/H
130 NEXT N
140 DRAW 0,L-N
```

---

14. Curves of pursuit in a circle

Choose the number of sectors
Choose the number of points per sector

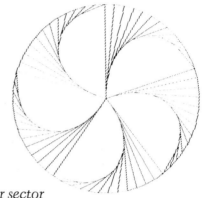

5 sectors
12 points per sector

B.B.C.

```
  0 REM Curves of Pursuit - Circle
 10 INPUT'"How many sectors? "V
 20 INPUT'"And points per sector? "N
 30 MODE 1
 40 GCOL 0,1
 50 MOVE 640,1000
 60 FOR B=0 TO 6.29 STEP PI/18
 70 DRAW 500*SIN B+640,500*COS B+500
 80 NEXT B
 90 GCOL 0,3
100 K=V*N
110 L=1000*PI/K
120 FOR B=0 TO 6.28 STEP 2*PI/V
130 X=0
140 Y=0
150 FOR T=B TO 6.29/V+B STEP 2*PI/K
160 P=500*SIN T
170 Q=500*COS T
180 MOVE X+640,Y+500
190 DRAW P+640,Q+500
200 H=SQR((P-X)^2+(Q-Y)^2)
210 X=X+L*(P-X)/H
220 Y=Y+L*(Q-Y)/H
230 NEXT T,B
```

Spectrum

```
 10 REM Pursuit Curves - Circle
 20 INPUT "How many sectors?";V
 30 INPUT "Points per sector";N
 40 CIRCLE INK 2,127,87,87
 50 LET K=2*PI/(V*N)
 60 LET L=87*K
 70 FOR B=0 TO 6.28 STEP 2*PI/V
 80 LET X=0
 90 LET Y=0
100 FOR T=B TO 6.29/V+B STEP K
110 LET P=87*SIN T-X
120 LET Q=87*COS T-Y
130 PLOT X+127,Y+87
140 DRAW P,Q
150 LET H=SQR (P*P+Q*Q)
160 LET X=X+L*P/H
170 LET Y=Y+L*Q/H
180 NEXT T
190 NEXT B
```

Programs

---

15. Tractrix

Choose the length of step

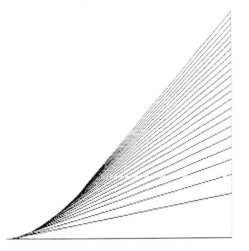

Step length 20

B.B.C.

```
  0 REM Tractrix
 10 INPUT'"Length of step? (4-100) "L
 20 MODE 1
 30 X=0
 40 Y=0
 50 FOR N=1 TO 420/L+1
 60 H=SQR(2200*X-X*X)
 70 MOVE X+100,Y
 80 DRAW 1200,Y+H
 90 X=X+L*(1100-X)/1100
100 Y=Y+L*H/1100
110 NEXT N
120 DRAW 1200,0
```

Spectrum

```
 10 REM Tractrix
 20 INPUT "Step length(1-30)";L
 30 LET X=0
 40 LET Y=0
 50 FOR N=0 TO 65/L
 60 LET H=SQR (400*X-X*X)
 70 PLOT X+25,Y
 80 DRAW 200-X,H
 90 LET X=X+L*(200-X)/200
100 LET Y=Y+L*H/200
110 NEXT N
120 DRAW 0,0.7*L-H-Y
```

---

16. Equal chords

Choose the number of points
Choose the number of steps

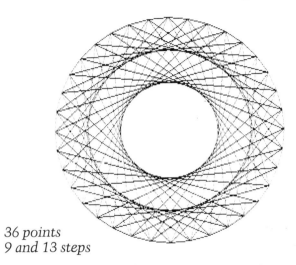

36 points
9 and 13 steps

B.B.C.

```
  0 REM Equal Chords
 10 INPUT'"How many points? "V
 20 MODE 1
 30 FOR T=3 TO 1 STEP -2
 40 PRINT TAB(11,0)SPC 6
 50 INPUT TAB(0,0)"And steps? "N
 60 GCOL 0,T
 70 MOVE 640,1000
 80 FOR H=0 TO 6.29 STEP PI/18
 90 DRAW 500*SIN H+640,500*COS H+500
100 NEXT H
110 FOR B=0 TO 6.28 STEP 2*PI/V
120 H=B+2*PI*N/V
130 MOVE 500*SIN B+640,500*COS B+500
140 DRAW 500*SIN H+640,500*COS H+500
150 NEXT B,T
```

Spectrum

```
 10 REM Equal Chords
 20 INPUT "How many points? ";V
 30 FOR T=1 TO 2
 40 INPUT "And steps? ";N
 50 CIRCLE INK T,127,87,87
 60 FOR B=0 TO 6.28 STEP 2*PI/V
 70 LET H=B+2*PI*N/V
 80 LET X=87*SIN B
 90 LET Y=87*COS B
100 PLOT X+127,Y+87
110 DRAW 87*SIN H-X,87*COS H-Y
120 NEXT B
130 NEXT T
```

Programs

---

17. Decreasing chords

Choose the number of points

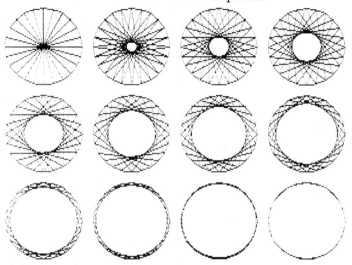

24 points (12 steps to 1 step)

B.B.C.

```
  0 REM Chords - Decreasing steps
 10 INPUT'"How many points? "V
 20 MODE 0
 30 N=INT(V/2)+1
 40 FOR D=340 TO 200 STEP -320
 50 FOR A=160 TO 1120 STEP 320
 60 IF N=1 THEN END ELSE N=N-1
 70 PRINT TAB(2,0);N" Steps "
 80 MOVE A,D+140
 90 FOR H=0 TO 6.29 STEP PI/18
100 DRAW 140*SIN H+A,140*COS H+D
110 NEXT H
120 FOR B=0 TO 6.28 STEP 2*PI/V
130 H=B+2*PI*N/V
140 MOVE 140*SIN B+A,140*COS B+D
150 DRAW 140*SIN H+A,140*COS H+D
160 NEXT B,A,D
```

Spectrum

```
 10 REM Chords - Decr. steps
 20 INPUT "How many points? ";V
 30 LET N=INT (V/2)+1
 40 FOR D=147 TO 29 STEP -59
 50 FOR A=42 TO 225 STEP 61
 60 IF N=1 THEN STOP
 70 LET N=N-1
 80 PRINT AT 0,0;N;" "
 90 CIRCLE A,D,28
100 FOR B=0 TO 6.28 STEP 2*PI/V
110 LET H=B+2*PI*N/V
120 LET X=28*SIN B
130 LET Y=28*COS B
140 PLOT X+A,Y+D
150 DRAW 28*SIN H-X,28*COS H-Y
160 NEXT B: NEXT A: NEXT D
```

---

18. Mystic rose

Choose the number of points

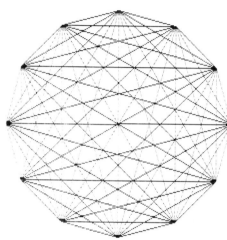

12 points

B.B.C.

```
  0 REM Mystic Rose
 10 INPUT'"How many points? "V
 20 MODE 1
 30 K=2*PI/V
 40 FOR B=K TO 6.28 STEP K
 50 FOR H=B+K TO 6.29 STEP K
 60 MOVE 500*SIN B+640,500*COS B+500
 70 DRAW 500*SIN H+640,500*COS H+500
 80 NEXT H,B
```

Spectrum

```
 10 REM Mystic Rose
 20 INPUT "How many points? ";V
 30 LET K=2*PI/V
 40 FOR B=K TO 6.28 STEP K
 50 FOR H=B+K TO 6.29 STEP K
 60 LET X=87*SIN B
 70 LET Y=87*COS B
 80 PLOT X+127,Y+87
 90 DRAW 87*SIN H-X,87*COS H-Y
100 NEXT H: NEXT B
```

Programs

19. Six mystic roses

Choose the least number of points

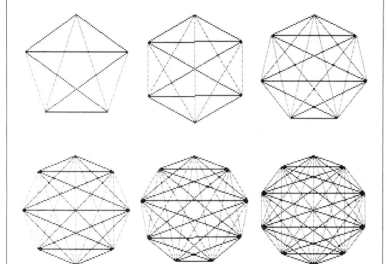

Starting with 5 points

B.B.C.

```
  0 REM 6 Mystic Roses
 10 INPUT'"Enter min points "V
 20 MODE 0
 30 FOR D=750 TO 250 STEP -500
 40 FOR A=200 TO 1076 STEP 438
 50 K=2*PI/V
 60 FOR B=K TO 6.28 STEP K
 70 FOR H=B+K TO 6.29 STEP K
 80 MOVE 200*SIN B+A,200*COS B+D
 90 DRAW 200*SIN H+A,200*COS H+D
100 NEXT H,B
110 V=V+1
120 NEXT A,D
```

Spectrum

```
 10 REM 6 Mystic Roses
 20 INPUT "Enter min points ";V
 30 FOR D=134 TO 41 STEP -93
 40 FOR A=40 TO 214 STEP 87
 50 LET K=2*PI/V
 60 FOR B=K TO 6.28 STEP K
 70 FOR H=B+K TO 6.29 STEP K
 80 LET X=40*SIN B
 90 LET Y=40*COS B
100 PLOT X+A,Y+D
110 DRAW 40*SIN H-X,40*COS H-Y
120 NEXT H: NEXT B
130 LET V=V+1
140 NEXT A: NEXT D
```

20. Concentric circles

Choose clockwise or anti-clockwise
Choose the points' ratio of the circles

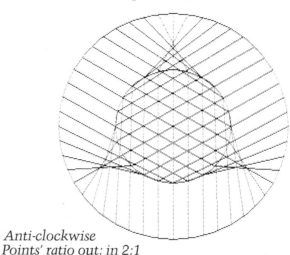

Anti-clockwise
Points' ratio out: in 2:1

B.B.C.

```
  0 REM Concentric Circles
 10 INPUT'"Key 1-CW or 0-ACW "V
 20 INPUT'"Points' ratio Out/In? "N
 30 MODE 1
 40 GCOL 0,1
 50 FOR T=252 TO 504 STEP 252
 60 MOVE 640,T+504
 70 FOR B=0 TO 6.29 STEP PI/18
 80 DRAW T*SIN B+640,T*COS B+504
 90 NEXT B,T
100 GCOL 0,3
110 FOR B=0 TO 6.28 STEP PI/(12*N)
120 IF V THEN H=N*B ELSE H=-N*B
130 MOVE 504*SIN B+640,504*COS B+504
140 DRAW 252*SIN H+640,252*COS H+504
150 NEXT B
```

Spectrum

```
 10 REM Concentric Circles
 20 INPUT "Key1-CW or 0-ACW ";V
 30 INPUT "Pts ratio Out/In ";N
 40 CIRCLE INK 2,127,87,43
 50 CIRCLE INK 2,127,87,86
 60 LET K=PI/(12*N)
 70 FOR B=0 TO 6.28 STEP K
 80 LET H=(2*V-1)*N*B
 90 LET X=86*SIN B
100 LET Y=86*COS B
110 PLOT X+127,Y+87
120 DRAW 43*SIN H-X,43*COS H-Y
130 NEXT B
```

Programs

---

21. Two linked circles

Choose the number of points on each circle
Choose the number of steps

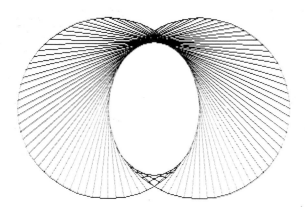

72 points
17 steps

B.B.C.

```
  0 REM 2 Linked Circles
 10 INPUT'"Points per circle? "V
 20 INPUT'"And how many steps? "N
 30 MODE 1
 40 GCOL 0,1
 50 FOR A=440 TO 840 STEP 400
 60 MOVE A,900
 70 FOR B=0 TO 6.29 STEP PI/18
 80 DRAW 400*SIN B+A,400*COS B+500
 90 NEXT B,A
100 GCOL 0,3
110 FOR B=0 TO 6.28 STEP 2*PI/V
120 H=B+2*PI*N/V
130 MOVE 400*SIN B+440,400*COS B+500
140 DRAW 400*SIN H+840,400*COS H+500
150 NEXT B
```

Spectrum

```
 10 REM 2 Linked Circles
 20 INPUT "Points per circle";V
 30 INPUT "And steps? ";N
 40 CIRCLE INK 2,85,87,85
 50 CIRCLE INK 2,170,87,85
 60 FOR B=0 TO 6.28 STEP 2*PI/V
 70 LET H=B+2*PI*N/V
 80 LET X=85*SIN B-85
 90 LET Y=85*COS 3
100 PLOT X+170,Y+87
110 DRAW 85*SIN H-X,85*COS H-Y
120 NEXT B
```

---

22. Three linked circles

Choose the number of points on each circle
Choose the number of steps

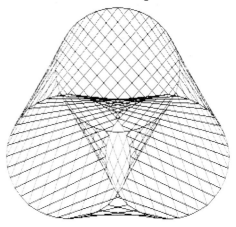

36 points
34 steps

B.B.C.

```
  0 REM 3 Linked Circles
 10 INPUT'"Points per circle? "V
 20 INPUT'"And how many steps? "N
 30 MODE 1
 40 GCOL 0,1
 50 FOR A=400 TO 880 STEP 240
 60 IF A=640 THEN D=420 ELSE D=0
 70 MOVE A,D+600
 80 FOR B=0 TO 6.29 STEP PI/18
 90 DRAW 300*SIN B+A,300*COS B+300+D
100 NEXT B,A
110 GCOL 0,3
120 FOR B=0 TO 6.28 STEP 2*PI/V
130 H=B+2*PI*N/V
140 G=B+4*PI*N/V
150 MOVE 300*SIN B+880,300*COS B+300
160 DRAW 300*SIN H+640,300*COS H+720
170 DRAW 300*SIN G+400,300*COS G+300
180 DRAW 300*SIN B+880,300*COS B+300
190 NEXT B
```

Spectrum

```
 10 REM 3 Linked Circles
 20 INPUT "Points per circle";V
 30 INPUT "And steps? ";N
 40 CIRCLE INK 2,87,52,52
 50 CIRCLE INK 2,128,123,52
 60 CIRCLE INK 2,169,52,52
 70 FOR B=0 TO 6.28 STEP 2*PI/V
 80 LET H=B+2*PI*N/V
 90 LET G=B+4*PI*N/V
100 LET P=52*SIN B
110 LET Q=52*COS B
120 LET R=52*SIN H
130 LET S=52*COS H
140 LET X=52*SIN G
150 LET Y=52*COS G
160 PLOT P+169,Q+52
170 DRAW R-P-41,S-Q+71
180 DRAW X-R-41,Y-S-71
190 DRAW P-X+82,Q-Y
200 NEXT B
```

Programs

---

23. Envelope of a roulette

Choose epicycloid or hypocyloid
Choose the number of cusps

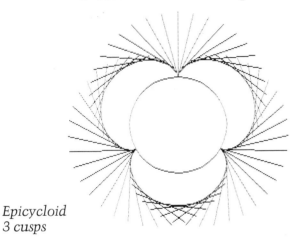

Epicycloid
3 cusps

B.B.C.

```
  0 REM Roulette - Envelope
 10 INPUT'"1-Epicyc or 0-Hypocyc "V
 20 INPUT'"How many cusps? "C
 30 MODE 1
 40 K=C/2+2*V
 50 G=500/K
 60 T=500-2*G*V
 70 GCOL 0,1
 80 MOVE 640,T+500
 90 FOR B=0 TO 6.29 STEP PI/18
100 DRAW T*SIN B+640,T*COS B+500
110 NEXT B
120 GCOL 0,3
130 FOR B=0 TO 6.28 STEP PI/(9*K)
140 H=2*PI*(1-V)-(1-2*V)*(B*K-B)
150 X=(500-G)*SIN B+640
160 Y=(500-G)*COS B+500
170 MOVE G*SIN H+X, G*COS H+Y
180 DRAW G*SIN(H+PI)+X, G*COS(H+PI)+Y
190 NEXT B
```

Spectrum

```
 10 REM Roulette - Envelope
 20 INPUT "1-Epicyc 0-Hypoc ";V
 30 INPUT "How many cusps? ";C
 40 LET K=C/2+2*V
 50 LET G=87/K
 60 LET T=87-2*G*V
 70 CIRCLE INK 2,127,87,T
 80 LET N=2*PI*(1-V)
 90 LET J=PI/(9*K)
100 FOR B=0 TO 6.28 STEP J
110 LET H=N-(1-2*V)*(B*K-B)
120 LET P=(87-G)*SIN B
130 LET Q=(87-G)*COS B
140 LET R=G*SIN H
150 LET S=G*COS H
160 LET X=G*SIN (H+PI)
170 LET Y=G*COS (H+PI)
180 PLOT R+P+127,S+Q+87
190 DRAW X-R,Y-S
200 NEXT B
```

---

24. Outline of a roulette

B.B.C.

```
  0 REM Roulette - Outline
 10 INPUT'"1-Epicyc or 0-Hypocyc "V
 20 INPUT'"How many cusps? "C'
 30 IF V=0 PRINT"Make turns < cusps"
 40 INPUT'"And turns? "N
 50 MODE 1
 60 IF V=0 AND C<=N THEN 30
 70 R=500/(C/N+2*V)
 80 G=R*(C/N+2*V-1)
 90 H=12*(C+2*V)/N^(1-3*V/4)
100 MOVE 640,G+(1-2*V)*R+508
110 FOR B=0 TO 6.28*N STEP PI/H
120 X=G*SIN B-R*SIN(B*G/R)
130 Y=G*COS B+(1-2*V)*R*COS(B*G/R)
140 DRAW X+640,Y+508
150 NEXT B
```

Spectrum

```
 10 REM Roulette - Outline
 20 INPUT "1-Epicyc 0-Hypoc ";V
 30 INPUT "How many cusps? ";C
 40 PRINT "If Hypo,turns<cusps"
 50 INPUT "And turns? ";N: CLS
 60 IF V=0 AND C<=N THEN RUN
 70 LET R=87/(C/N+2*V)
 80 LET G=R*(C/N+2*V-1)
 90 LET H=6*(C+2*V)/(N-N*V+V)
100 LET K=R*(1-2*V)
110 LET P=0: LET Q=G+K
120 FOR B=0 TO 6.28*N STEP PI/H
130 LET T=B*G/R
140 LET X=INT (G*SIN B-R*SIN T)
150 LET Y=INT (G*COS B+K*COS T)
160 PLOT P+127,Q+88
170 DRAW X-P,Y-Q
180 LET P=X: LET Q=Y
190 NEXT B
```

---

25. Numbering of a roulette

B.B.C.

```
  0 REM Roulette - Numbering
 10 @%=3
 20 INPUT'"1-Epicyc 0-Hypocyc "V
 30 INPUT'"How many cusps? "C
 40 INPUT'"And points on circle? "N
 50 IF V THEN K=C+1 ELSE K=C-1
 60 MODE 7
 70 FOR B=0 TO N-1
 80 H=V+(K*(B+1)-V) MOD N
 90 PRINT TAB(8*(B DIV20),B MOD20);
100 IF V=1 THEN PRINT B+1,H
110 IF V=0 THEN PRINT N-H,B+1
120 NEXT B
```

Spectrum

```
 10 REM Roulette - Numbering
 20 INPUT "1-Epicyc 0-Hypoc ";V
 30 INPUT "How many cusps? ";C
 40 INPUT "Points on circle?";N
 50 LET K=C+2*V-1
 60 FOR B=1 TO N
 70 LET H=K*B-INT ((K*B-V)/N)*N
 80 LET D=B-22*INT ((B-1)/22)
 90 LET A=6*INT ((B-1)/22)
100 PRINT AT D-1,A;
110 IF V=1 THEN PRINT B;"-";H
120 IF V=0 THEN PRINT N-H;"-";B
130 NEXT B
```

26. Epicycloid

Choose the number of cusps
Choose the number of points on the circle

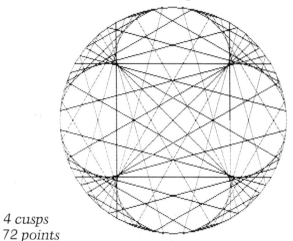

4 cusps
72 points

B.B.C.

```
  0 REM Epicycloid
 10 INPUT'"How many cusps? "C
 20 INPUT'"And points on circle? "N
 30 MODE 1
 40 GCOL 0,1
 50 MOVE 776,1000
 60 FOR H=0 TO 6.29 STEP PI/18
 70 DRAW 500*SIN H+776,500*COS H+500
 80 NEXT H
 90 GCOL 0,3
100 T=0
110 FOR B=0 TO 6.28 STEP 2*PI/N
120 PRINT TAB(T DIV 30*3,T MOD 30);
130 PRINT;1+(C*T+C+T) MOD N;
140 T=T+1
150 H=B*(C+1)
160 MOVE 500*SIN B+776,500*COS B+500
170 DRAW 500*SIN H+776,500*COS H+500
180 NEXT B
```

Spectrum

```
 10 REM Epicycloid
 20 INPUT'"How many cusps? ";C
 30 INPUT "Points on circle?";N
 40 CIRCLE INK 2,167,87,87
 50 LET T=0
 60 FOR B=0 TO 6.28 STEP 2*PI/N
 70 LET G=(C+1)*(T+1)
 80 LET R=G-INT ((G-1)/N)*N
 90 LET D=T-22*INT (T/22)
100 LET A=3*INT (T/22)
110 PRINT AT D,A;R
120 LET T=T+1
130 LET H=B*(C+1)
140 LET X=87*SIN B
150 LET Y=87*COS B
160 PLOT X+167,Y+87
170 DRAW 87*SIN H-X,87*COS H-Y
180 NEXT B
```

27. Six epicycloids

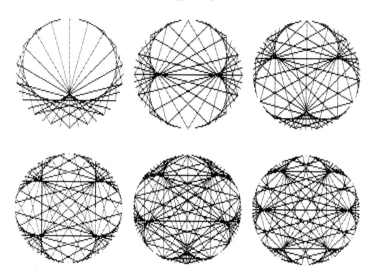

1–6 cusps

B.B.C.

```
  0 REM 6 Epicycloids
 10 MODE 0
 20 C=1
 30 FOR D=750 TO 250 STEP -500
 40 FOR A=200 TO 1072 STEP 436
 50 MOVE A,D+200
 60 FOR H=0 TO 6.29 STEP PI/18
 70 DRAW 200*SIN H+A,200*COS H+D
 80 NEXT H
 90 K=6*(C+2)
100 FOR B=0 TO 6.28 STEP PI/K
110 H=B*(C+1)
120 MOVE 200*SIN B+A,200*COS B+D
130 DRAW 200*SIN H+A,200*COS H+D
140 NEXT B
150 C=C+1
160 NEXT A,D
```

Spectrum

```
 10 REM 6 Epicycloids
 20 LET C=1
 30 FOR D=134 TO 41 STEP -93
 40 FOR A=40 TO 214 STEP 87
 50 CIRCLE A,D,41
 60 LET K=PI/(6*(C+2))
 70 FOR B=0 TO 6.28 STEP K
 80 LET H=B*(C+1)
 90 LET X=40*SIN B
100 LET Y=40*COS B
110 PLOT X+A,Y+D
120 DRAW 40*SIN H-X,40*COS H-Y
130 NEXT B
140 LET C=C+1
150 NEXT A: NEXT D
```

Programs

---

28. Cardiod from circles

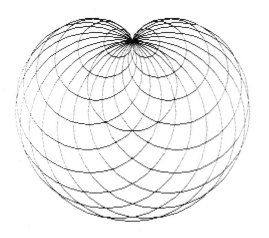

B.B.C.

```
  0 REM Cardioid - Circles
 10 MODE 1
 20 GCOL 0,1
 30 MOVE 640,848
 40 FOR B=0 TO 6.29 STEP PI/18
 50 DRAW 200*SIN B+640,200*COS B+648
 60 NEXT B
 70 GCOL 0,3
 80 FOR T=PI/10 TO 1.9*PI STEP PI/10
 90 X=200*SIN T
100 Y=200*COS T
110 H=SQR(X^2+(200-Y)^2)
120 MOVE X+640,Y+H+648
130 FOR B=0 TO 6.29 STEP PI/18
140 DRAW H*SIN B+X+640,H*COS B+Y+648
150 NEXT B,T
```

Spectrum

```
10 REM Cardioid - Circles
20 CIRCLE INK 2,127,117,39
30 FOR T=0 TO 2*PI STEP PI/10
40 LET X=39*SIN T
50 LET Y=39*COS T
60 LET H=SQR (X*X+(39-Y)^2)
70 CIRCLE X+127,Y+117,H
80 NEXT T
```

---

29. Nephroid from circles

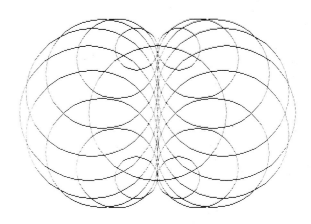

B.B.C.

```
  0 REM Nephroid - Circles
 10 MODE 1
 20 GCOL 0,1
 30 MOVE 640,800
 40 FOR B=0 TO 6.29 STEP PI/18
 50 DRAW 300*SIN B+640,300*COS B+500
 60 NEXT B
 70 GCOL 0,3
 80 FOR H=PI/10 TO 1.1*PI STEP PI
 90 FOR T=H TO H+PI/1.2 STEP PI/10
100 X=300*SIN T
110 Y=300*COS T+500
120 MOVE X+640,X+Y
130 FOR B=0 TO 6.29 STEP PI/18
140 DRAW X*SIN B+X+640,X*COS B+Y
150 NEXT B,T,H
```

Spectrum

```
10 REM Nephroid - Circles
20 CIRCLE INK 2,127,87,62
30 FOR T=0 TO 2*PI STEP PI/10
40 LET X=62*SIN T
50 LET Y=62*COS T
60 CIRCLE X+127,Y+87,X
70 NEXT T
```

Programs

30. Cycloid

Choose the number of arches
Choose the number of steps per arch

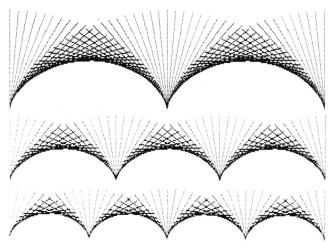

Arches/steps per arch 2/36, 3/24, 4/18

B.B.C.

```
  0 REM Cycloid
 10 INPUT'"How many arches? "V
 20 INPUT'"And steps per arch? "N
 30 MODE 1
 40 T=400/V
 50 FOR B=0 TO 3.15*V STEP PI/N
 60 H=B+PI
 70 MOVE T*SIN B+T*B,T*COS B+500
 80 DRAW T*SIN H+T*B,T*COS H+500
 90 NEXT B
```

Spectrum

```
 10 REM Cycloid
 20 INPUT "How many arches? ";V
 30 INPUT "Steps per arch? ";N
 40 LET T=81/V
 50 FOR B=0 TO 3.15*V STEP PI/N
 60 LET H=B+PI
 70 LET X=T*SIN B
 80 LET Y=T*COS B
 90 PLOT X+T*B,Y+87
100 DRAW T*SIN H-X,T*COS H-Y
110 NEXT B
```

31. Hypocycloid

Choose the number of cusps
Choose the number of points per cusp

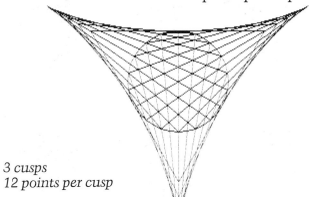

3 cusps
12 points per cusp

B.B.C.

```
  0 REM Hypocycloid
 10 INPUT'"How many cusps? "C
 20 INPUT'"And points per cusp? "N
 30 MODE 1
 40 GCOL 0,1:MOVE 680,904
 50 FOR B=0 TO 6.29 STEP PI/18
 60 DRAW 224*SIN B+680,224*COS B+680
 70 NEXT B:GCOL 0,3
 80 K=C*N:T=0
 90 FOR B=0 TO 6.28 STEP 2*PI/C
100 G=B+PI/C
110 FOR H=B TO 6.29/C+B STEP 2*PI/K
120 PRINT TAB(T DIV 24*3,T MOD 24+6);
130 PRINT;K-(C-1)*(T+1) MOD K;
140 IF T<K-1 THEN T=T+1
150 J=-H*(C-1)
160 M=COS(H*C/2-H+G)
170 L=224*COS(-H*C/2)/(M-(M=0))
180 IF ABS M<1E-5 THEN L=224*C/(C-2)
190 MOVE 224*SIN J+680,224*COS J+680
200 DRAW L*SIN G+680,L*COS G+680
210 NEXT H,B
```

Spectrum

```
 10 REM Hypocycloid
 20 INPUT "How many cusps? ";C
 30 INPUT "Points per cusp? ";N
 40 CIRCLE INK 2,154,116,38
 50 LET K=C*N: LET F=2*PI/K
 60 LET T=0: LET V=38*C/(C-2)
 70 FOR B=0 TO 6.28 STEP 2*PI/C
 80 FOR H=B TO 6.29/C+B STEP F
 90 LET G=K-(C-1)*(T+1)
100 LET R=G-INT ((G-1)/K)*K
110 LET D=T-18*INT (T/18)
120 LET A=3*INT (T/18)
130 PRINT AT D+4,A;R
140 IF T<K-1 THEN LET T=T+1
150 LET M=COS (H*C/2-H+B+PI/C)
160 LET S=M-(M=0): LET J=H-H*C
170 LET L=38*COS (-H*C/2)/S
180 IF ABS M<1E-5 THEN LET L=V
190 LET P=38*SIN J+154
200 LET Q=38*COS J+116
210 LET X=L*SIN (B+PI/C)+154
220 LET Y=L*COS (B+PI/C)+116
230 PLOT P,Q: DRAW X-P,Y-Q
240 NEXT H: NEXT B
```

Programs

32. Six hypocycloids

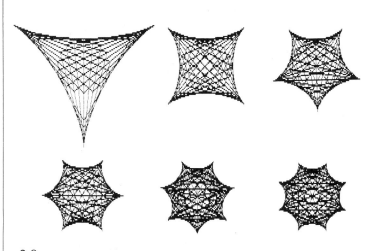

3-8 cusps

B.B.C.

```
  0 REM 6 Hypocycloids
 10 MODE 0
 20 C=3
 30 FOR D=750 TO 250 STEP -500
 40 FOR A=260 TO 1108 STEP 424
 50 FOR B=0 TO 6.28 STEP 2*PI/C
 60 G=B+PI/C
 70 FOR H=B TO 6.29/C+B STEP PI/(6*C)
 80 J=-H*(C-1)
 90 M=COS(H*C/2-H+G)
100 L=100*COS(-H*C/2)/(M-(M=0))
110 IF ABS M<1E-5 THEN L=100*C/(C-2)
120 MOVE 100*SIN J+A,100*COS J+D
130 DRAW L*SIN G+A,L*COS G+D
140 NEXT H,B
150 C=C+1
160 NEXT A,D
```

Spectrum

```
 10 REM 6 Hypocycloids
 20 LET C=3
 30 FOR D=135 TO 40 STEP -95
 40 FOR A=53 TO 223 STEP 85
 50 LET F=PI/(3*C)
 60 LET V=20*C/(C-2)
 70 FOR B=0 TO 6.28 STEP 2*PI/C
 80 FOR H=B TO 6.29/C+B STEP F
 90 LET M=COS (H*C/2-H+B+PI/C)
100 LET S=M-(M=0): LET J=H-H*C
110 LET L=20*COS (-H*C/2)/S
120 IF ABS M<1E-5 THEN LET L=V
130 LET P=20*SIN J+A
140 LET Q=20*COS J+D
150 LET X=L*SIN (B+PI/C)+A
160 LET Y=L*COS (B+PI/C)+D
170 PLOT P,Q: DRAW X-P,Y-Q
180 NEXT H: NEXT B: LET C=C+1
190 NEXT A: NEXT D
```

33. Astroid

Choose the number of steps

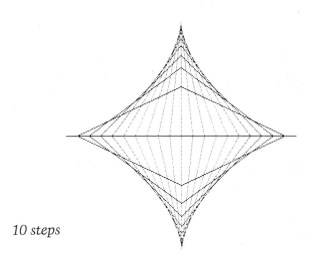

10 steps

B.B.C.

```
  0 REM Astroid
 10 INPUT'"How many steps? "V
 20 MODE 2
 30 VDU 23,1,0;0;0;0;
 40 T=V*(500 DIV V)
 50 REPEAT GCOL 0,RND(7)
 60 FOR N=0 TO T STEP 500 DIV V
 70 H=SQR(T^2-N^2)
 80 MOVE 640+N,T
 90 DRAW 640,T+H
100 DRAW 640-N,T
110 DRAW 640,T-H
120 DRAW 640+N,T
130 NEXT N
140 UNTIL INKEY(100)>0
```

Spectrum

```
 10 REM Astroid
 20 INPUT "How many steps? ";V
 30 INK INT (RND*7)
 40 FOR N=0 TO 85 STEP 84/V
 50 LET H=SQR ABS (84^2-N^2)
 60 PLOT N+127,87
 70 DRAW -N,H
 80 DRAW -N,-H
 90 DRAW N,-H
100 DRAW N,H
110 NEXT N
120 PAUSE 30
130 GO TO 30
```

Programs

34. Four curves

Parabola *Pursuit*

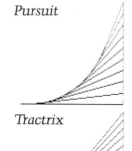

¼ Astroid *Tractrix*

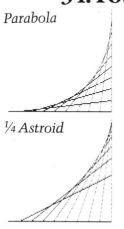

B.B.C.

```
  0 REM 4 Curves
 10 MODE 1
 20 PRINT TAB(2,2)"Parabola"
 30 FOR N=100 TO 532 STEP 48
 40 MOVE N,500:DRAW 532,N+400
 50 NEXT N
 60 PRINT TAB(1,20)"1/4 Astroid"
 70 FOR N=0 TO 432 STEP 48
 80 MOVE N+100,0
 90 DRAW 532,SQR(864*N-N*N)
100 NEXT N
110 PRINT TAB(22,2)"Pursuit"
120 L=48:X=748:Y=500
130 FOR N=500 TO 932 STEP 48
140 MOVE X,Y:DRAW 1200,N
150 H=SQR((1200-X)^2+(N-Y)^2)
160 X=X+L*(1200-X)/H:Y=Y+L*(N-Y)/H
170 NEXT N:DRAW 1200,500
180 PRINT TAB(22,20)"Tractrix"'
190 L=23:X=0:Y=0
200 FOR N=1 TO 9:H=SQR(896*X-X*X)
210 MOVE X+748,Y:DRAW 1200,Y+H
220 X=X+L*(448-X)/448:Y=Y+L*H/448
230 NEXT N:DRAW 1200,0
```

Spectrum

```
 10 REM 4 Curves
 20 PRINT AT 2,2;"Parabola"
 30 FOR N=46 TO 127 STEP 9
 40 PLOT N,94: DRAW 127-N,N-46
 50 NEXT N
 60 PRINT AT 13,2;"1/4 Astroid"
 70 FOR N=0 TO 84 STEP 7
 80 LET H=SQR (168*N-N*N)
 90 PLOT N+43,0: DRAW 84-N,H
100 NEXT N
110 PRINT AT 2,19;"Pursuit"
120 LET X=171: LET Y=94
130 FOR N=94 TO 175 STEP 9
140 PLOT X,Y: DRAW 255-X,N-Y
150 LET H=(255-X)^2+(N-Y)^2
160 LET R=SQR H: LET L=8
170 LET X=X+L*(255-X)/R
180 LET Y=Y+L*(N-Y)/R
190 NEXT N: DRAW 0,-81
200 PRINT AT 13,19;"Tractrix"
210 LET L=9: LET X=0: LET Y=0
220 FOR N=1 TO 5
230 LET H=SQR (168*X-X*X)
240 PLOT X+171,Y: DRAW 84-X,H
250 LET X=X+L*(84-X)/84
260 LET Y=Y+L*H/84
270 NEXT N: DRAW 0,-81
```

35. Spiral

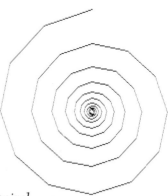

Equiangular spiral
30° 10 turns

B.B.C.

```
  0 REM Spiral of Archimedes &
 10 REM Equiangular Spiral
 20 INPUT'"Key 1-Arch. or 0-Equi. "V
 30 INPUT'"Radial line length?(mm) "R
 40 INPUT'"Enter angle (5-45) "N
 50 INPUT'"Turns?(less than angle)";T
 60 IF T>N THEN 50
 70 INPUT'"Init. dist. from centre "H
 80 IF V=0 AND H=0 THEN 70
 90 J=360*T/N
100 IF V=1 THEN G=(R-H)/J
110 IF V=0 THEN G=(R/H)^(1/J)
120 MODE 1
130 MOVE 820,500+450*H/R
140 FOR L=0 TO J
150 PRINT TAB(L DIV 30*3,L MOD 30);
160 PRINT;INT(H+0.5);
170 K=450*H/R
180 B=L*RAD N
190 DRAW K*SIN B+820,K*COS B+500
200 IF V THEN H=H+G ELSE H=H*G
210 NEXT L
```

Spectrum

```
 10 REM Spiral of Archimedes &
 20 REM Equiangular Spiral
 30 INPUT "1-Arch. 0-Equi. ";V
 40 INPUT "Radial line? mm ";R
 50 INPUT "Angle? (5-45) ";N
 60 INPUT "Turns? ( < angle)";T
 70 IF T>N THEN GO TO 60
 80 INPUT "Init. dist? mm ";H
 90 IF V=2*(H>0) THEN GO TO 80
100 LET J=360*T/N: LET F=1/J
110 LET T=PI*N/180
120 IF V=1 THEN LET G=(R-H)/J
130 IF V=0 THEN LET G=(R/H)^F
140 LET P=169: LET Q=87*(H/R+1)
150 FOR L=0 TO J: LET M=L
160 IF L>241 THEN LET M=241
170 LET D=M-22*INT (M/22)
180 LET A=3*INT (M/22)
190 PRINT AT D,A;INT (H+0.5)
200 LET B=L*T: LET K=87*H/R
210 LET X=INT (K*SIN B)+169
220 LET Y=INT (K*COS B)+87
230 PLOT P,Q: DRAW X-P,Y-Q
240 LET H=(H+G)*V+H*G*(V=0)
250 LET P=X: LET Q=Y: NEXT L
```

6.
Proofs and demonstrations

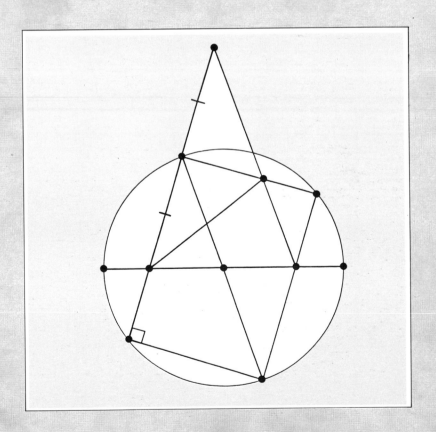

Proofs and Demonstrations

$$ax^2 + 2hxy$$

$$+ by^2 + 2gx$$

$$+ 2fy + c$$

$$= 0$$

In the earlier chapters of this book, the sewing rules for each pattern or curve were presented simply as rules to follow. By following the instructions accurately, the desired curve was generated. In this chapter the intention is to go further and to prove that each rule does in fact generate the curve in question. It is easy enough to recognise a circle, but we need to know, for instance, that a "parabola" is in fact a true parabola and not some curve which is just "parabola-like".

Each stitch that is sewn is a straight line, and the patterns can be regarded as the successive positions of a line moving as it obeys certain mathematical rules. The curve which is traced out is known as an envelope and each stitch only contributes a small portion to it. In theory it only contributes a single point, the point of contact between the tangent and the curve. This point of contact may move along the line as the line itself moves. To find the locus of the point which traces out the envelope as the line moves according to different rules demands a variety of techniques.

A general theory of curve stitching requires a general theory of envelopes and much work with partial differentiation. Having obtained an algebraic equation for the envelope, there remains the difficulty of recognising exactly what it represents. Even simple curves like circles and parabolas are hard to recognise in general algebraic form. There are of course tests which can be applied to determine with certainty exactly what type of curve a particular equation represents, and some readers may find a great deal of interest and enjoyment in such work. However it is not really appropriate for a book of this level and we shall take a more specific approach. We shall attempt to find some particular property of a curve, or some particularly convenient set of co-ordinates which will enable us to prove that the curve is indeed the one we say it is.

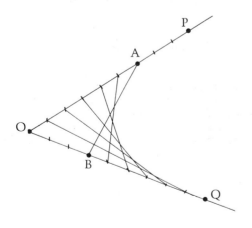

Sewing a parabola

Let us begin by proving the standard construction which we have used for parabolas in many different patterns earlier in the book. The standard method was to mark along one axis OP the same number of equal divisions as along another axis OQ. Then each stitch was sewn between points A on OP and B on OQ chosen so that A was as many divisions in from P as B was from O. The equal divisions on one axis do not have to be the same length as the equal divisions on the other.

This proportional relationship is true when the standard construction is followed:

$$\frac{OA}{OP} = \frac{BQ}{OQ}$$

We now need to prove that if this proportional relationship is true, then every possible position of AB is a tangent to the same parabola.

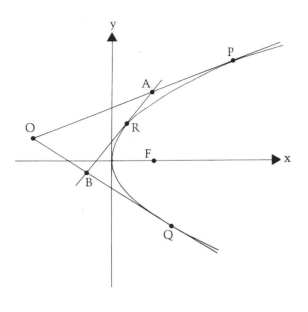

Let us use the simplest algebraic form of the parabola which is $y^2 = 4ax$. For it to be expressed in this form the x-axis has to be the axis of symmetry and the y-axis tangential to the parabola. The focus is then $F(a, O)$. Points on the parabola can be expressed in parametric form.

Let us consider two tangents at $P(ap^2, 2ap)$ and $Q(aq^2, 2aq)$ meeting at O, which we will now prove is the point $(apq, a(p+q))$.

The equation of OP is $py = x + ap^2$ and of OQ is $qy = x + aq^2$

So $y = \dfrac{x}{p} + ap$ and $y = \dfrac{x}{q} + aq$

Giving $\dfrac{x}{p} - \dfrac{x}{q} = aq - ap$

Therefore $x = apq$

Substituting this value in $py = x + ap^2$ produces $y = a(p + q)$.

Likewise if AB is a tangent to the parabola at $R(ar^2, 2ar)$, then A is $(apr, a(p + r))$ and B is $(aqr, a(q + r))$.

We can now find both ratios using the x co-ordinates only.

$$\dfrac{OA}{OP} = \dfrac{apr - apq}{ap^2 - apq} = \dfrac{r - q}{p - q}$$

$$\dfrac{BQ}{OQ} = \dfrac{aq^2 - aqr}{aq^2 - apq} = \dfrac{q - r}{q - p}$$

Therefore it is true that $\qquad \dfrac{OA}{OP} = \dfrac{BQ}{OQ}$

and so the tangent to the parabola does divide the axes in the same way as the stitches are sewn using the standard construction. It also shows that if there are (say) 10 divisions on each axis, then the parabola touches each axis at the 11th.

Curves which are parabola-like

Once the basic construction of the parabola is mastered and some simple patterns have been sewn, there is a natural inclination to apply the same technique to different starting situations. This can be a most creative activity and the number of beautiful designs which can be produced is virtually limitless. From a strictly mathematical viewpoint it is best to be aware that the curves which are generated are 'parabola-like', rather than true parabolas. Only if the proportion rule given opposite is true does the procedure give a parabola.

This illustration, which is reproduced in colour on page 35, shows the kind of pattern which can be sewn. The proportion rule is not obeyed, but eight of these 'pseudo-parabolas' do combine very attractively to give a 'four leaved clover' design within a circle.

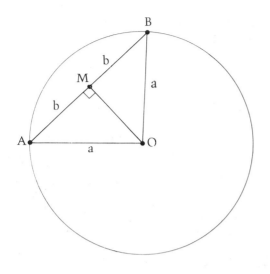

Equal chords of a circle

Several of the curve stitching patterns make use of equal chords of a circle to generate another smaller concentric circle. It is easy to see why this is true.

For all chords

$$OM^2 = a^2 - b^2$$

where a is the radius of the circle and b is half the length of the chord.

Hence OM is always the same length for equal chords and thus M lies on a circle centre O. M is the point of contact between the tangent AB on the envelope it traces out.

Some equal chord patterns are what is called unicursal, which means they can be drawn without lifting the pencil from the paper, as long as the number of points on the circle and the number of steps around the circle are relatively prime, meaning that they have no common factor. For the purpose of curve stitching in the usual way this property has no significance but if, however, instead of holes through the card, the points were eyelets on the surface or nails knocked into a base board, then those designs which are unicursal could be made with a single piece of thread or wire.

The pattern in six colours

On page 23 there is an equal chord pattern, sewn in six colours. This table shows which points to join to obtain it.

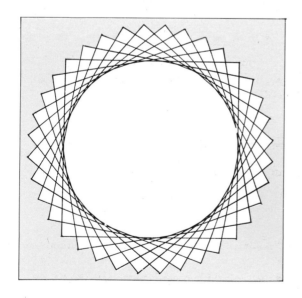

GREEN	RED	YELLOW	MAUVE	CYAN	ORANGE
20 - 29	knot 27 - 36	14 - 5	knot 3 - 30	26 - 17	knot 15 - 6
19 - 28	28 - 1	13 - 4	4 - 31	25 - 16	16 - 7
18 - 27	29 - 2	12 - 3	5 - 32	24 - 15	17 - 8
20 - 11	9 - 36	14 - 23	21 - 30	26 - 35	33 - 6
19 - 10	10 - 1	13 - 22	22 - 31	25 - 34	34 - 7
18 - 9	knot 11 - 2	12 - 21	knot 23 - 32	24 - 33	knot 35 - 8

The shading indicates those stitches which are on the back of the card. Complete all the stitches in one colour before knotting it to the other colour of the pair. Red and green are knotted together twice as are yellow and mauve and then cyan and orange.

76

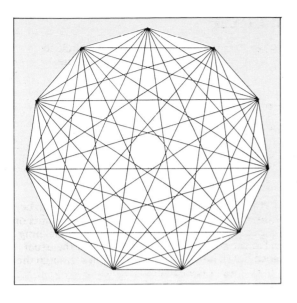

Mystic roses

A mystic rose is the name given to patterns which are made where the circumference of a circle is marked into equal divisions and then all the points are joined to all the others.

POINTS	2	3	4	5	6	7	8	9	10	n
DIAGONALS	1	3	6	10	15	21	28	36	45	$\frac{1}{2}n(n-1)$
REGIONS	2	4	8	16	30					

Sewing these kinds of patterns may well stimulate an interest in drawing mystic roses and in investigating their properties. They make an interesting topic for investigation because while the number of diagonals is clearly related to the number of points the number of regions is not. For mystic roses of 2, 3, 4, 5 points, the number of regions is 2, 4, 8, 16 respectively and a clear pattern appears to be established. However this is upset by the fact that a mystic rose of 6 points divides the circle into 30 regions, not 32. Further work is needed.

An ellipse from inverses

The pattern on page 39 generates an ellipse from two parallel lines simply by joining each number on one line to its inverse on another. To prove that this is truly an ellipse is not difficult.

If an ellipse is referred to its major and minor axes as axes of co-ordinates, its equation is

$$\frac{x^2}{a^2} + \frac{y^2}{b^2} = 1$$

The tangents at the ends of the major axis are then $x = a$ and $x = -a$.

There are two tangents to an ellipse with the same gradient, one each side of the centre. If the gradient is m, the equations are

$$y = mx + \sqrt{a^2m^2 + b^2} \qquad \text{and} \qquad y = mx - \sqrt{a^2m^2 + b^2}$$

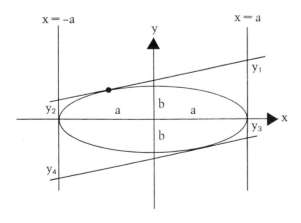

when $\quad x = a \quad y_1 = \ am + \sqrt{a^2m^2 + b^2} \quad$ and $\quad y_3 = \ am - \sqrt{a^2m^2 + b^2}$

when $\quad x = -a \quad y_2 = -am + \sqrt{a^2m^2 + b^2} \quad$ and $\quad y_4 = -am - \sqrt{a^2m^2 + b^2}$

hence $\quad y_1y_2 = a^2m^2 + (b^2 - a^2m^2) \qquad$ and $\qquad y_3y_4 = a^2m^2 + (b^2 - a^2m^2)$

$$= \ b^2 \qquad\qquad\qquad\qquad = b^2$$

The product of the distances from the x axis is b^2. If b is chosen to be 1, then the points to join are inverses.

This is exactly the curve stitching method which is used to generate the ellipse.

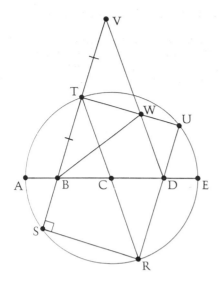

An ellipse from chords

The method on page 40 which generated an ellipse by using a set-square in a circle is also not difficult to prove, this time using elementary geometry.

Suppose the circle has radius r and that AE is a diameter. The points B and D which are to be the foci of the ellipse are equally spaced from A and E respectively. Let us imagine a rectangle RSTU on the circle so that the sides TS and UR pass through B and D. Then TR passes through C, the centre of the circle.

Produce ST to V so that BT = TV. Join VD to intersect TU at W. Join BW.

Since VT is equal and parallel to DR, VDRT is a parallelogram, so VD = TR = 2r
The triangles BWT and VWT are congruent and so BW = WV
Hence BW + WD = VW + WD = VD = TR = 2r.

This is the well known focal property of the ellipse.

This curve stitching construction means that the sewn stitches are the sides TU of the rectangles which pass through B and D. On each position of TU there is a single point W which lies on the ellipse. W is therefore the point of contact of the tangent with the ellipse and therefore lies on the envelope.

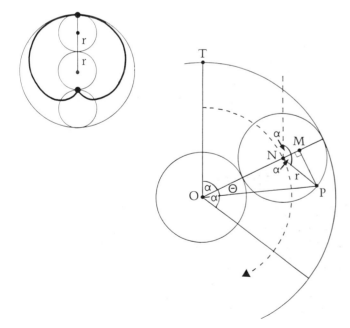

Epicycloids – the cardioid

The pattern on page 44 illustrates the envelope of a cardioid. It was generated by sewing between each point and its double around the circle. There was also an explanation that a cardioid is the path traced out by a point on a circle as it rolls around another circle with the same radius. We now need to prove that these two different points of view are in fact different descriptions of the same curve.

Let us find the position of a point P on the cardioid.

A point on the circumference of the rolling circle will have turned through 2α as ON turns through α. The position of the point P can be described by the length of OP and the angle $(\alpha + \Theta)$.

$NM = r \cos \alpha$, $PM = r \sin \alpha$.

$$OP^2 = OM^2 + PM^2$$
$$= (2r + r \cos \alpha)^2 + (r \sin \alpha)^2$$
$$= 5r^2 + 4r^2 \cos \alpha$$

Also $\sin \Theta = \dfrac{PM}{OP} = \dfrac{r \sin \alpha}{OP}$

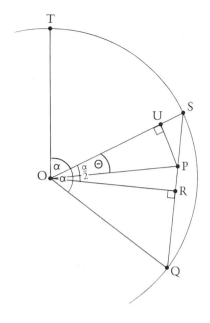

Now let us look at the problem from a curve stitching point of view.

Each stitch joins two points, one of which is twice as far around the circle from this starting point as the other. If T is the starting point and TÔS is α, then TÔQ is 2α. We need to show that there is a single point P on SQ which contributes to the cardioid. We test the point P which is one third of the way along SQ. If the choice of one third seems arbitrary, we say that it was suggested by the fact that the cusp of the cardioid is one third of the way along the diameter.

Since the radius of the sewing circle is 3r, the length of $SR = 3r \sin \frac{\alpha}{2}$ and $PR = r \sin \frac{\alpha}{2}$

$$OP^2 = OR^2 + PR^2$$

$$= (3r \cos \frac{\alpha}{2})^2 + (r \sin \frac{\alpha}{2})^2$$

$$= r^2 + 8r^2 \cos^2 \frac{\alpha}{2}$$

$$= r^2 + 4r^2 (1 + \cos \alpha)$$

$$= 5r^2 + 4r^2 \cos \alpha$$

Since $PU = SP \cos \frac{\alpha}{2}$, we have

$$\sin \Theta = \frac{2r \sin \frac{\alpha}{2} \cos \frac{\alpha}{2}}{OP} = \frac{r \sin \alpha}{OP}$$

Hence the point P obtained by this method is the same point P as that obtained by the other method. This confirms that the method of curve stitching which is given on page 44 does produce the envelope of a cardioid.

The reader might now care to prove the similar results for the nephroid on page 45, where P is one quarter of the way along SQ, and for the epicycloid of Cremona on page 46 where P is one fifth of the way along SQ. In each case find the length of OP and the angle Θ by both methods.

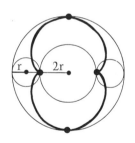
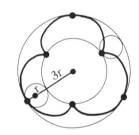

Hypocycloids – the deltoid

On page 48 there is a pattern which generates the envelope of a deltoid by sewing between pairs of points, one on a circle and one on a line leading to the cusp. Each stitch joins a point to another point which is twice as far round the circle, but in the opposite direction. Unlike the epicycloids where both end-points lie on the circle, each stitch is sewn through one point and then passes over the other, forming the envelope outside the initial circle. There is also a diagram which shows that a deltoid is the path traced out by a point on the circumference of a circle as it rolls around the inside of a circle with three times the radius. We now need to show that these two different points of view are in fact describing the same curve.

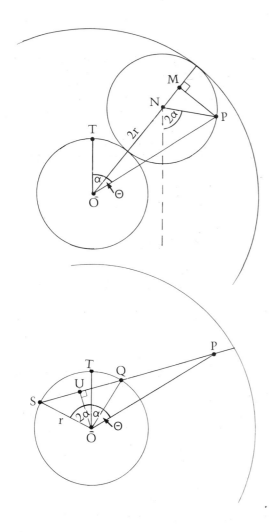

Let us find the position of a point P on the deltoid.

P is a point on the circumference of the rolling circle. As ON rotates clockwise through α, PN rotates anticlockwise through 2α. The position of P can be described by the length of OP and the angle $\alpha + \Theta$.

$$PM = r \sin 3\alpha, \quad OM = 2r + r \cos(180 - 3\alpha) = 2r - r \cos 3\alpha$$

$$OP^2 = PM^2 + OM^2$$

$$= (r \sin 3\alpha)^2 + (2r - r \cos 3\alpha)^2$$

$$= 5r^2 - 4r^2 \cos 3\alpha$$

Also $\sin \Theta = \dfrac{PM}{OP} = \dfrac{r \sin 3\alpha}{OP}$

Now let us look at the problem from a curve stitching point of view. Each stitch joins two points, one of which is twice as far around the circle as the starting point, but in the opposite direction. If T is the starting point and $T\hat{O}Q$ is α, then $T\hat{O}S = 2\alpha$.

We need to show that there is a single point P on SQ produced which contributes to the deltoid. We test the point P which is such that PQ = QS, suggested by the fact that the cusp of the deltoid extends beyond the inner circle by the length of that circle's diameter.

$$OU = r \cos \frac{3\alpha}{2}, \quad UP = 3r \sin \frac{3\alpha}{2}$$

$$OP^2 = OU^2 + UP^2$$

$$= (r \cos \frac{3\alpha}{2})^2 + (3r \sin \frac{3\alpha}{2})^2$$

$$= r^2 + 8r^2 \sin^2 \frac{3\alpha}{2}$$

$$= r^2 + 4r^2 (1 - \cos 3\alpha)$$

$$= 5r^2 - 4r^2 \cos 3\alpha$$

To find an expression for Θ, we describe the area of the triangle OQP in two different ways.

$$PQ = SQ = 2r \sin \frac{3\alpha}{2}$$

$$\tfrac{1}{2} r \, OP. \sin \Theta = \tfrac{1}{2} PQ. OU = \tfrac{1}{2} 2r \sin \frac{3\alpha}{2} \cos \frac{3\alpha}{2}$$

$$\text{therefore } \sin \Theta = \frac{r \sin 3\alpha}{OP}$$

Hence the point P which is obtained by this method is the same point as that obtained by the first method. This confirms that the curve stitching method on page 48 does generate the envelope of a deltoid.

The reader might care to prove similar results for the astroid on page 50 where PQ = ½ SQ and for the six-cusped hypocloid on page 52 where PQ = ¼ SQ. In each case find the length of OP and the angle Θ by both methods.

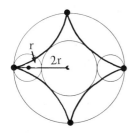
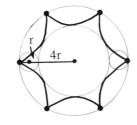

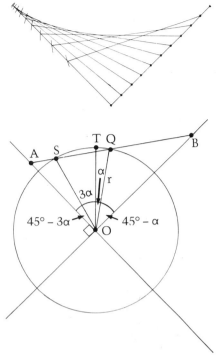

The Astroid

On page 51 there is a very simple method of sewing an astroid which seems to have no connection with that shown on page 50 or indeed with the fact that it is a hypocycloid. The method suggested is to sew a series of stitches of constant length so that their ends lie on a pair of straight lines at right angles to each other. This can be described as the "sliding ladder" problem. To prove it is fortunately a simple exercise in elementary geometry.

Points S, T and Q are on a circle of radius r. If Q is rotated α clockwise from T, then S is rotated through 3α anticlockwise from T. The chord SQ meets lines through O which are at 45° to the line OT at points A and B.

SOQ is an isosceles triangle and hence $S\hat{Q}O = 90 - 2\alpha$

$Q\hat{O}B = 45° - \alpha$. Hence $Q\hat{B}O = 45° - \alpha$ and thus OQB is isosceles.

Hence QB = QO = r.

$Q\hat{O}A = 45° + \alpha$. Hence $Q\hat{A}O = 45° + \alpha$ and thus OQA is isosceles.

Hence QA = QO = r

Thus AB = AQ + QB = 2r which is independent of α.

This proves that the astroid is the envelope of lines of constant length with their ends on two lines at right angles.

Stitching spirals

A spiral is a curve traced out by a moving point which systematically increases or decreases its distance from a fixed point at the centre while turning round it. There are two main types which can be sewn, Archimedian and equiangular. In each case the stitches suggest the curve by means of consecutive line segments, a quite different approach from the normal technique of curve stitching where each stitch only contributes a small portion to the curve.

Archimedian spiral

An Archimedian spiral is one in which the radial distance increases in direct proportion to the angle turned. A simple example of it is a coiled rope.

In order to sew an Archimedian spiral like the one illustrated here it is necessary to measure and mark out on the back of the card every point which is to be sewn. In this example there are 10 turns and 12 radii spaced at 30° intervals and it is convenient to mark it out by first drawing 10 concentric circles with radii increasing in steps of 6mm. Then at each radial line mark a point ½mm further out. After 12 radii the spiral has moved out by 6mm and just reaches the next concentric circle. Having pierced all the marked points, the thread is sewn in two stages. Firstly from the outside alternately towards the centre and then back out again filling the gaps. It is best to avoid those holes which would come within the first circle as they are too close together. "Cheat" a little by joining the centre to a point on the first circle with radius 6mm.

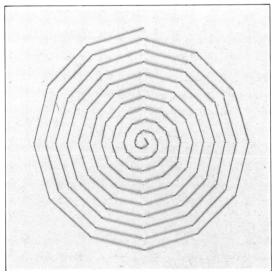

Equiangular spiral

This spiral is so called because it cuts each of the radial lines at equal angles. It is widely observed in nature, for instance the shell of a nautilus or other molluscs. Flowers such as the daisy and sunflower have heads which contain a series of intersecting equiangular spirals.

The diagram shows that each line segment of the spiral meets the next radius at the same angle, forming a sequence of similar triangles of increasing size.

Along each of the evenly spaced radii of an Archimedian spiral the points are spaced as an arithmetic progression. Along each radius of an equiangular spiral they are spaced as a geometric progression. The next distance is obtained by multiplying the previous distance by a number which is called the common ratio.

To sew the equiangular spiral in the illustration the common ratio was chosen to be 1.035. The first three distances were 1, 1 x 1.035, and 1 x $(1.035)^2$. When all the distances are calculated and then measured out on the back of the card, the holes can be pierced. It is sensible to round off each distance to the nearest ½mm.

This table gives the calculation for 10 turns. The ones at the centre were so close together that it was practicable to sew only the outer seven turns.

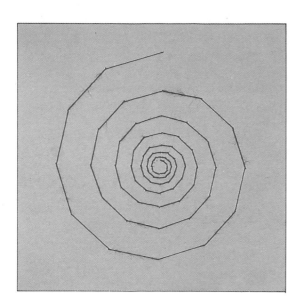

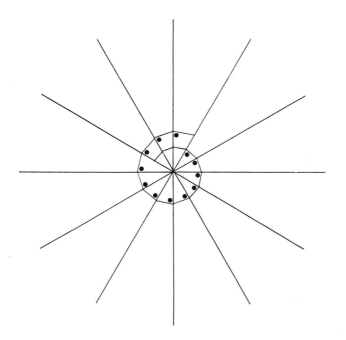

Distances in mm from the centre along each radial line.

360	1									
30	1	1.5	2.5	3.5	5.5	8	12	18	27.5	41
60	1	1.5	2.5	3.5	5.5	8.5	12.5	19	28.5	42.5
90	1	1.5	2.5	4	5.5	8.5	13	19.5	29.5	44
120	1	1.5	2.5	4	6	9	13.5	20	30.5	45.5
150	1	2	2.5	4	6	9	14	21	31.5	47.5
180	1	2	3	4	6.5	9.5	14.5	21.5	32.5	49
210	1.5	2	3	4.5	6.5	10	15	22.5	33.5	50.5
240	1.5	2	3	4.5	7	10	15.5	23	35	52.5
270	1.5	2	3	4.5	7	10.5	16	24	36	54
300	1.5	2	3	5	7	11	16.5	24.5	37	56
330	1.5	2	3.5	5	7.5	11. 5	17	25.5	38.5	58
360	1.5	2.5	3.5	5	7.5	11.5	17.5	26.5	40	60

The sewing is done in two stages, from the outside alternately in towards the centre and then back out again filling in the gaps.

7.
Curve Stitching in three dimensions

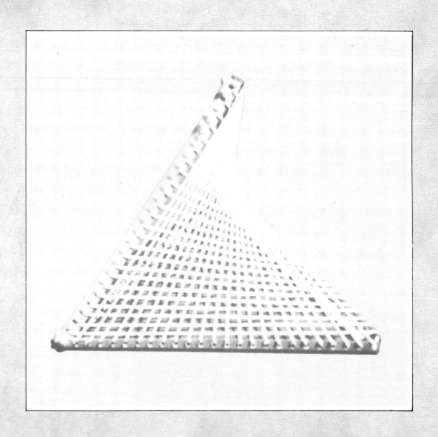

Working in three dimensions

Only a few lines of approach are suggested in this chapter as it would take at least another book to cover the endless possibilities which are offered by three dimensional curve stitching. The ideas on this page are essentially mathematical, but on the pages which follow are some examples of how curve stitching can be regarded as a branch of sculpture, interpreted by four well-known twentieth century artists.

Some kind of framework is required which is strong enough to withstand the tensions of the strings. For the very smallest designs, card may be sufficiently rigid, but usually metal or wood is needed. Perforated strips of Meccano work well and they can be bolted together into their required positions in space. For small scale constructions button thread will suffice, but coloured wool or string will show up better for anything larger.

Some constructions use what is essentially two dimensional curve stitching, but repeated in several different planes. A simple example of this approach is the "Parabola Tree" shown in Figure 1. A central upright pole is fixed to a base made from cross-pieces. A standard parabola pattern is then sewn between the upright and each of the cross-pieces. The stringing for each cross-piece then lies in a single plane. The diagram shows four-cross pieces with only two having been sewn, but there could be any number from three upwards.

A different approach is shown in Figure 2, where the strings are sewn between two pairs of opposite sides of a tetrahedron. A fifth strut is needed to keep the construction rigid. In this case the strings form a curved surface in three dimensional space. Since all the strings are straight, such surfaces are known as "ruled" surfaces. This one is called a hyperbolic paraboloid and is a shape which has been used for roofs. The strings are replaced by straight iron reinforcing rods, and the roof is then cast from concrete to cover them.

Figure 3 shows a framework which is in the form of a cube and there are many different ways of sewing between the sides. The diagram shows the two dimensional kind of envelope where the strings join adjacent sides and the three dimensional kind of ruled surface where the strings join non-adjacent sides.

Another approach is shown in Figure 4. The strings join the edges of two circles which have been cut out of a pair of hinged cards. Button thread will do very well as there is no need for it to be very large. Alternatively, try shirring elastic and see how the envelope changes as the angle between the card is altered.

A simplified version of the Lagrange Stringed Mathematical Model of 1872 which is in the Science Museum in London is well worth constructing. Two discs of strong card are centrally mounted at opposite ends of a spindle which allows the discs to rotate. They are kept apart by a tube on the spindle. Make holes near the edge of each card and thread shirring elastic through the corresponding holes to form a cylindrical cage. Then hold one disc while turning the other to make some very interesting effects. It is also possible to use a slide projector to project lines and curves on to the strings.

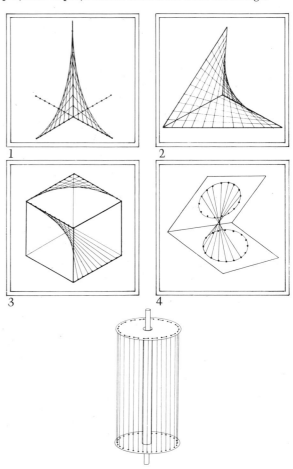

84

Curve stitching in art

Among the first to bring together mathematics and art through curve stitching was the Russian sculptor Naum Gabo (1890-1977). He was one of the founders of the Constructivist movement whose ideals were laid down in *The Realistic Manifesto* which he wrote with his brother Antoine Pevsner to accompany an exhibition of their work in Moscow in 1920.

In conventional sculpture, the object is usually created by carving away material from, say, wood or stone, to reveal a shape within the block. By contrast, the Constructive approach was to build up a design from separate parts.

By this time Gabo had developed strong views about the direction in which art, particularly sculpture, should be going and in the *Manifesto* he stated that "in sculpture the mass as a sculptural element" should be renounced. He further advocated bringing "back to sculpture the lines as a direction and in it we affirm a depth as the one form of space". Although mathematical influences are discernable in most of Gabo's work, the idea of suggesting volume and mass by enclosing space with strings did not emerge until the 1930s.

Antoine Pevsner's sculptures, while following the tenets propounded in the *Manifesto* are totally different from those of his brother. Pevsner incorporated curve stitching ideas by welding straight rods so closely together that they amounted to a curved surface. His *Developable Column* of 1942 exemplifies this technique. One later work that could truly be regarded as three-dimensional curve stitching is the *Maquette of a Monument Symbolizing the Liberation of the Spirit* dating from 1952 and now in the Tate Gallery.

In 1935 Gabo was persuaded to come to England where he joined the Hampstead artistic colony which included Henry Moore and Barbara Hepworth. Both had been influenced by Gabo's Constructivist ideas, and indeed Gabo had already known Barbara Hepworth and her husband Ben Nicholson in Paris.

Gabo had experimented with strings in sculpture before producing his *Spheric Theme 2nd Variation* in 1937-8 and it was at just this time that Henry Moore incorporated strings in some of his work, only to abandon them soon afterwards. The actual material for stringing varied; bronze wire and piano strings were used as was, a little later, nylon. On the other hand, it was not until 1938-9 that Barbara Hepworth followed Henry Moore's footsteps and began to include strings in her sculptures. She continued to do so after her move, towards the end of 1939, to St Ives in Cornwall where the potter Bernard Leach was already working, and some of these sculptures derive from her drawings of curves suggested by envelopes.

Soon, Gabo joined Barbara Hepworth and the other artists who were gathering in St Ives to escape wartime London, but in 1946 moved on to the United States where he spent his last thirty years. Many of his sculptures rely on his imaginative use of three-dimensional curve stitching for their remarkable effect. Among these are *Linear Construction No. 1, Linear Construction No. 2* and *Torsion, Variation*. The first two of these achieve a particular brilliance because their frameworks are made of Perspex which he started using in 1937 and nylon stringing added extra lustre. Often he produced several versions of his sculptures in different sizes. What, then, do strings of whatever material contribute to sculpture? They are undoubtedly central to Gabo's work; it would not be too much to say for many of them that the stringing is the sculpture. With Moore and Hepworth strings play a subtler role, and contrast is certainly an element when their tautness is compared with the smooth, rounded form of the mass itself. Strings can add a sense of weightlessness while guiding the eye from one part of the sculpture to another; at times they also add an extra depth.

This brief survey is limited to four sculptors who have responded to the possibilities inherent in curve stitching. However, many other artists have created works bearing little similarity to each other but nevertheless embodying the technique of using strings in two dimensions as well as three.

One such artist is Sue Fuller whose plastic thread construction *String Composition 128* was completed in 1964. Kenneth Martin's *Screw Mobile (1956-9)* in phosphor bronze and steel derives from related sources as do several of Richard Lippold's sculptures including *Variation No. 7 Full Moon (1949-50)* using nickel-chrome and stainless steel wire and brass rods, and his *1963 Flight* for the Pan-Am building in New York.

Naum Gabo

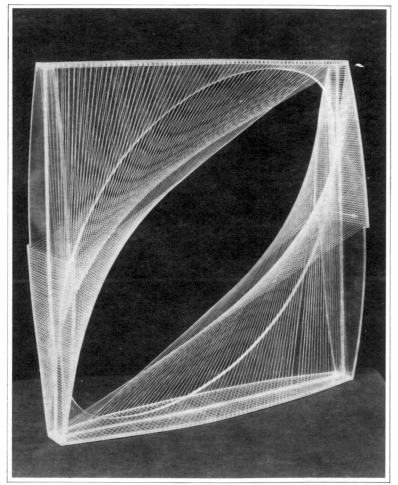

Linear Construction No. 1 (1942-3)
Tate Gallery, London
© Nina and Graham Williams 1989

Naum Gabo

Linear Construction No. 2 (1970-71)
Tate Gallery, London
© Nina and Graham Williams 1989

Naum Gabo

Antoine Pevsner   Naum Gabo

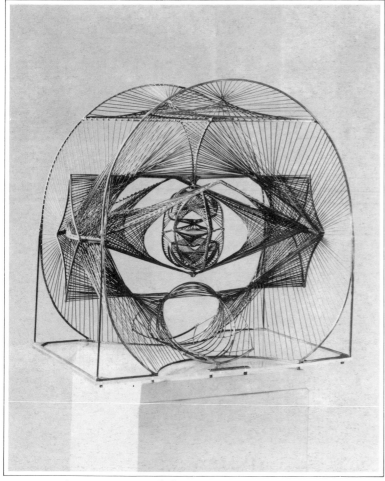

Maquette of a Monument Symbolising
the Liberation of the Spirit (1952)
Tate Gallery, London
© ADAGP Paris, DACS London 1989

Antoine Pevsner

Torsion, Variation (1963)
Nina Williams Collection
© Nina and Graham Williams 1989

Naum Gabo

Henry Moore

The Bride (1939-40)
Museum of Modern Art, New York
© Henry Moore Foundation 1989

Henry Moore

Stringed Figure (1938/60)
Tate Gallery, London
© Henry Moore Foundation 1989

Henry Moore

Barbara Hepworth

Winged Figure (1962)
John Lewis Building, London
© *Sir Alan Bowness 1989*

Barbara Hepworth

Curved Form (Bryher II) (1961)
Hirshhorn Sculpture Garden, Washington D.C.
© *Sir Alan Bowness 1989*

Barbara Hepworth

Bibliography

For further reading, the following sources are suggested. The author gratefully acknowledges his indebtedness to many of them for the valuable ideas they contain.

Mathematics

Archibald, R.C., Curves, Special. *Encyclopaedia Britannica*. 14th edn. 1929
Boltyanskii, V.C., *Envelopes*. Pergamon Press. 1964
Boole, M.E., *Collected Works*. C.W. Daniel. 4 vols. 1931
Boon, R.C., *A Companion to Elementary School Mathematics*. Longmans. 1924
Catranides, Peter, Inductive Geometry Through Curve-Stitching. *Mathematics Teaching No. 78*, March 1977
Catranides, Peter, Beyond the Parabolic Envelope. *Ibid No. 85*, December 1978
Cobham, E.M., *Mary Everest Boole - A Memoir*. C.W. Daniel. 1951
Cundy, H.M and Rollett, A.P., *Mathematical Models*. Tarquin Publications 3rd edn. 1981
Dorrie, Heinrich, Transl. David Austin. *100 Great Problems of Elementary Mathematics*. Dover 1965
Fielker, David S., *Mathematics and Curve Stitching*. Mathematics Teaching No. 64, September 1973
Gardner, Martin, Mathematical Games. *Scientific American*. February 1961 (Ellipse), July 1964 (Cycloid) and September 1970 (Roulettes)
James, E.J., *Curve Stitching*. Oxford University Press. 1960
Johnson, Donvan A, *Curves*. John Murray. 1966
Land, Frank, *The Language of Mathematics*. John Murray. 1960
Leapfrogs, *Curves*. Tarquin Publications. 1982
Lewis, K., *The Ellipse*. Longmans. 1969
Lockwood, E.H., *A Book of Curves*. Cambridge University Press. 1961
Madachy, Joseph, *Mathematics on Vacation*. Nelson. 1968
Mathematical Pie 25 (October 1958), *29* (February 1960), *46* (October 1965), *65* (Spring 1972)
Newton, Herbert C., ed., *Harmonic Vibrations and Vibration Figures*. Newton. 1909
Somervell, Edith L., *A Rhythmic Approach to Mathematics*. George Philip. 1906
Steinhaus, H., *Mathematical Snapshots*. 3rd American. edn. O.U.P. 1983
Tahta, D.G., *A Boolean Anthology*. Association of Teachers of Mathematics. 1972
Todd, Audrey, *The Maths Club*. Hamish Hamilton. 1968
Yates, Robert C., *Curves and their Properties*. National Council of Teachers of Mathematics. 3rd edn. 1974

Art

Bowness, Alan, intro., *Naum Gabo. Sixty Years of Constructivism*. Catalogue of an exhibition at the Tate Gallery. 1987
Compton, Michael, *Optical and Kinetic Art*. Tate Gallery. 1967
Hammacher, A.M., *Barbara Hepworth*. Thames and Hudson. 1968
Hodin, J.P., *Barbara Hepworth. Life and Work*. Lund Humphries. 1961
Read, Herbert, intro., *Naum Gabo/Antoine Pevsner*. Catalogue of an exhibition at the Museum of Modern Art, New York. 1948
Read, Herbert, *Henry Moore*. Thames and Hudson. 1965
Rickey, George, *Constructivism*. Studio Vista. 1968
Rogers, L.R., *Sculpture*. Oxford University Press. 1969

Proportional scale for 36 point circular templates

There is a 36 point circular template to cut out on page 93. If you prefer not to cut it out, or want one of a different size it is easy to make one using this diagram.

1. On a suitable piece of card or paper draw a circle of the size you want your template to be. Mark the centre of the circle and then cut around the circumference.

Place the centre of your circle on the centre of the proportional scale and mark off the number of points you require. This scale can be used for 36, 18, 12, 9, 6, 4 or 3 equally spaced points. With only a little estimating it can be used for 72 and 24 also.

2. If you want to mark points around an inner circle as well draw such a circle on your template. If the paper is not too thick you may be able to see lines of the proportional scale through it. If that is not possible then draw the radii to the points marked around the circumference. However, it is probably easier to make another smaller template with the diameter you require.

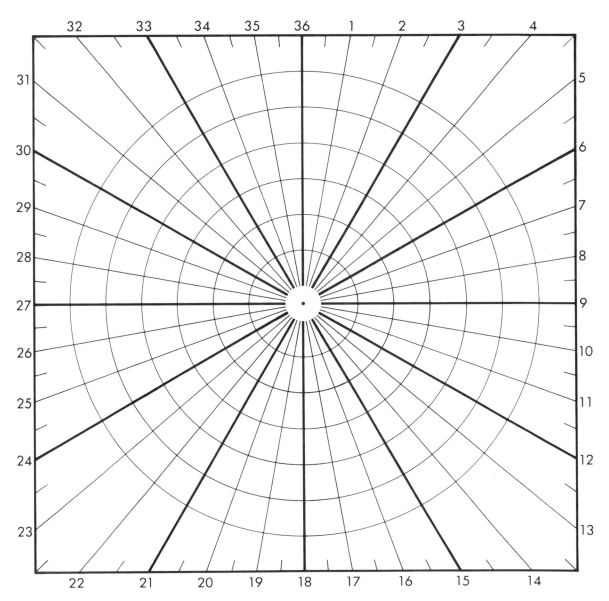

Proportional scale for 40 point circular templates

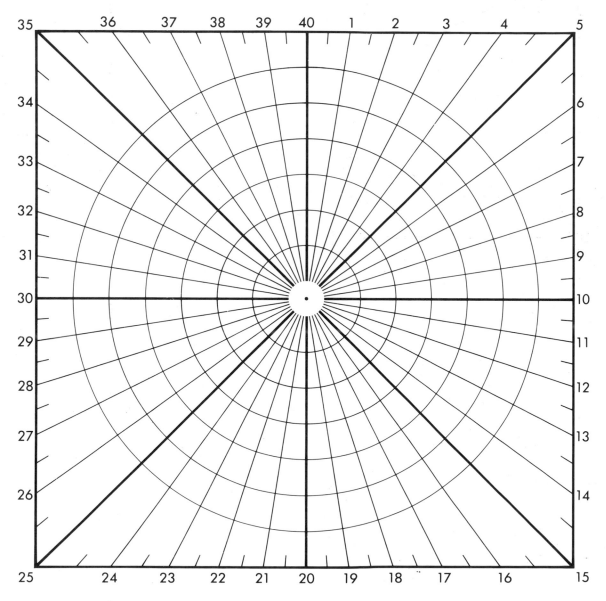

There is a 40 point circular template to cut out on page 95. If you prefer not to cut it out, or want one of a different size it is easy to make one using this diagram.

1. On a suitable piece of card or paper draw a circle of the size that you want your template to be. Mark the centre of the circle and then cut around the circumference.

Place the centre of your circle on the centre of the proportional scale and mark off the number of points you require. This scale can be used for 40, 20, 10, 8, 5 or 4 equally spaced points. With only a little estimating it can be used for 80 and 16 points also.

2. If you want to mark points around an inner circle as well draw such a circle on your template. If the paper is not too thick you may be able to see the lines of the proportional scale through it. If that is not possible then draw the radii to the points marked around the circumference. However, it is probably easier to make another smaller template with the diameter you require.

Cut along this line

36 point circular template

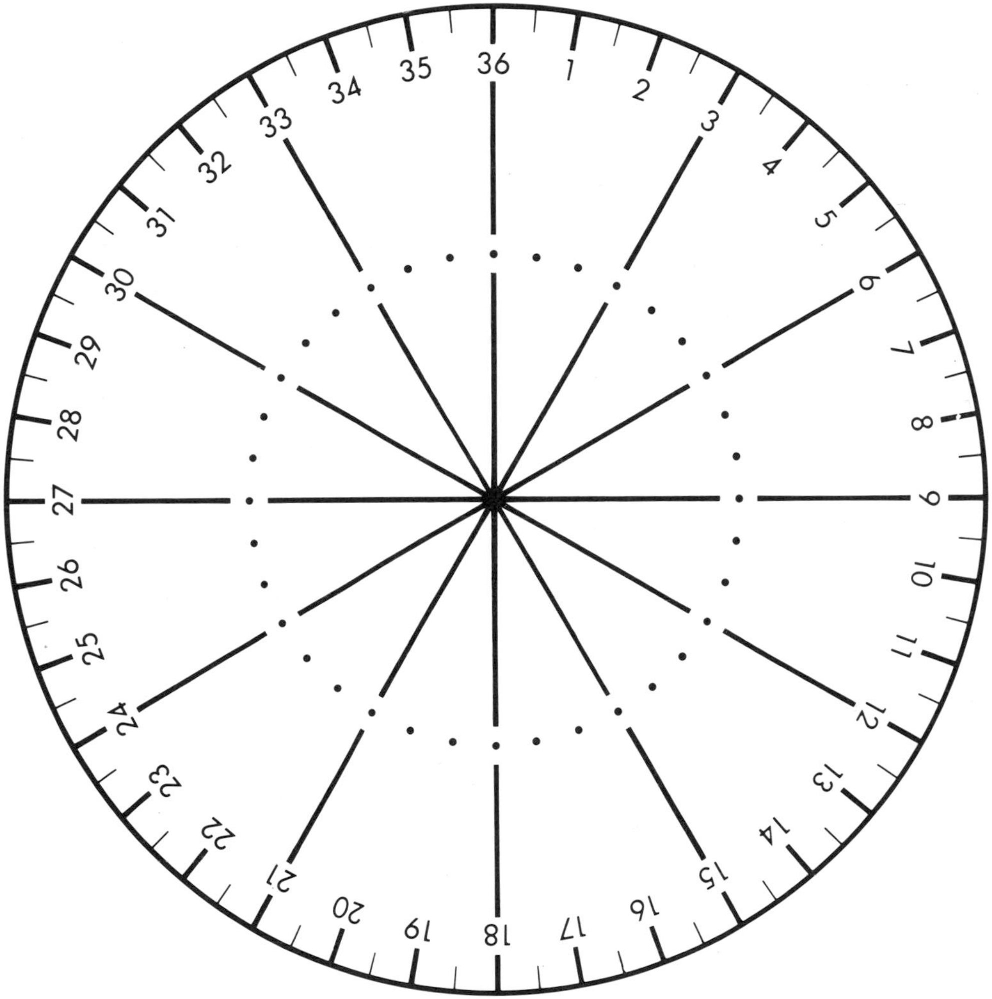

Cut out this circle and use it as a template to divide the circumference of a circle (radius 65mm) into 72, 36, 24, 18, 12, 9, 8, 6, 4 or 3 equal divisions, or to find the corners of a regular dodecagon, nonagon, hexagon, square or triangle.

The inner circle is half the diameter of the outer one.

Line division template

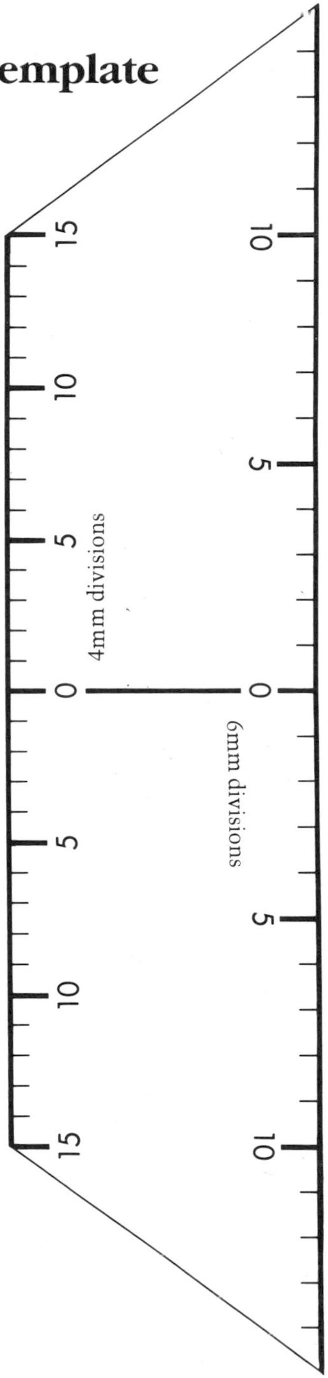

Cut out this scale and use it as a template to mark a line with 4 or 6mm divisions.

36 point circle
Outer Radius: 65 mm
Inner Radius: 32.5mm

40 point circular template

Cut along this line

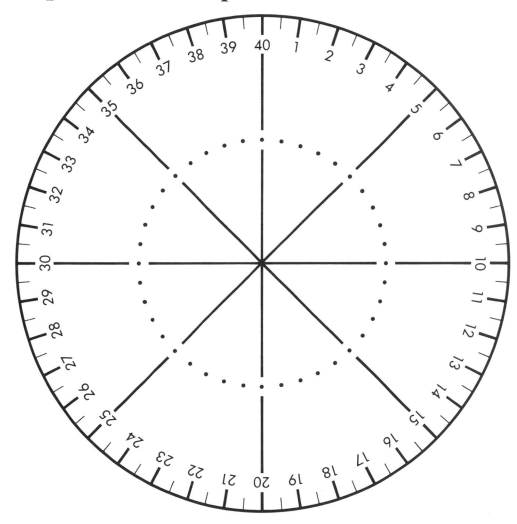

Cut out this circle and use it as a template to divide the circumference of a circle (radium 65mm) into 80, 40, 20, 16, 10, 8 or 4 equal divisions, or to find the corners of a regular, decagon, octagon or square.

The inner circle is half the diameter of the outer one.

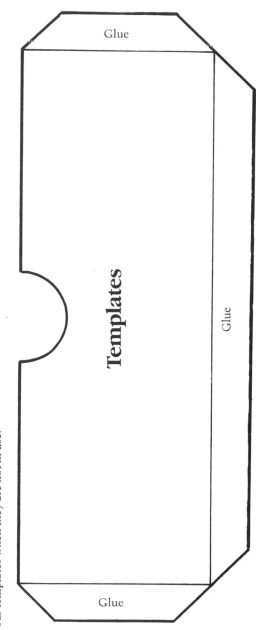

Keep your templates in here
Cut out this pocket, score and fold along the lines and then glue it inside the back cover to hold your templates when they are not in use.

Templates

Glue

Glue

Glue

40 point circle
Outer Radius: 65mm
Inner Radius: 32.5mm